Essays in Biochemistry

Other recent titles in the Essays in Biochemistry series:

Essays in Biochemistry volume 36: Molecular Trafficking
edited by P. Bernstein
2000 ISBN 1 85578 131 X

Essays in Biochemistry volume 35: Molecular Motors
edited by G. Banting and S.J. Higgins
2000 ISBN 1 85578 103 4

Essays in Biochemistry volume 34: Metalloproteins
edited by D.P. Ballou
1999 ISBN 1 85578 106 9

Essays in Biochemistry volume 33: Molecular Biology of the Brain
edited by S.J. Higgins
1998 ISBN 1 85578 086 0

Essays in Biochemistry volume 32: Cell Signalling
edited by D. Bowles
1997 ISBN 1 85578 071 2

volume 37 2001

Essays in Biochemistry

Regulation of Gene Expression

Edited by K.E. Chapman and S.J. Higgins

Portland Press

Essays in Biochemistry is published by Portland Press Ltd
on behalf of the Biochemical Society

Portland Press
59 Portland Place
London W1B 1QW, U.K.
Fax: 020 7323 1136;
e-mail: editorial@portlandpress.com
www.portlandpress.com

British Library Cataloguing-in-Publication Data
A catalogue record for this book is available from the British Library

ISBN   1 85578 138 7
ISSN   0071 1365

Typeset by Portland Press Ltd
Printed in Great Britain by Information Press Ltd, Eynsham, U.K.

Contents

6 Signalling from the cell surface to the nucleus
Melanie Lee and Stephen Goodbourn

7 Control of gene expression and the cell cycle
Ho Man Chan, Noriko Shikama and Nicholas B. La Thangue

8 Regulation of mRNA translation
Christopher G. Proud

9   To live or to die — a cell's choice
Martyn Link and David J. Harrison

10   Perspective
Nick Hastie

Preface

In all organisms, correct development, growth and function depends on the precise and exquisite control of the expression of the many thousands of genes comprising their genomes. The human genome is now estimated to contain around 30 000–40 000 genes and even the 'simple' bacterium, *Escherichia coli*, is encoded in about 4300 genes. Thanks to the heroic efforts of the Human Genome Project and other recent sequencing *tours de force* we now know the complete sequences of nearly all the genes of many different prokaryotic and eukaryotic organisms. However, it is not immediately apparent from these gene sequences how expression is regulated during development and in response to changes in the environment. Elucidating the mechanisms that control the processes leading to protein synthesis in the correct cell, at the precise time and in response to appropriate signals remains a major challenge in this 'post-genomic' era. The answers will not only be relevant to our understanding of the basic processes of gene expression, but will also be of immense practical importance in the application of gene therapy in medicine and the development of novel organisms for agriculture, for example.

The authors of the essays in this volume, all internationally recognized experts in their fields, have addressed many topical questions in gene regulation. The volume starts with a reiteration by Mark Ptashne and Alex Gann of a unifying hypothesis of transcriptional regulation by specific 'locators' or signals, frequently proteins (transcription factors), on the DNA which assist in the location of RNA polymerase(s) to the sites where transcription is to be initiated. More detailed descriptions of the process of transcription initiation and of the proteins that regulate it are given in the subsequent essays by Georgina Lloyd, Paolo Landini and Steve Busby and by Grace Gill. In recent years it has become clear that chromatin plays a fundamental and dynamic role in gene regulation and that alterations in chromatin structure can equip cells with a heritable 'memory' of the control state of particular genes. Hence, in his essay, Alan Wolffe discusses the roles of nucleosome structure and histone modification in the control of transcription, and Richard Meehan and Irina Stancheva provide an excellent illustration of the effects of the methylation state of DNA on gene transcription. Gene expression is a dynamic process that must be regulated in accordance with the needs of the organism in a constantly changing environment and according to the proliferative state of the cell. Melanie Lee and Stephen Goodbourn describe the mechanisms by which extracellular signals are transmitted to the nucleus, resulting in altered gene transcription, and Nick La Thangue and his colleagues review how transcriptional activity is coupled to progression through the cell cycle. Although the regulation of transcription is generally the most significant step in modulating the expression of a gene, control of protein synthesis can also occur at subsequent post-transcriptional stages, including RNA processing and mRNA translation. In this area, Chris Proud provides us with an overview of translational processes that regulate gene expression. Many of the areas highlighted in these earlier essays are touched on by Martyn Link and David Harrison when they describe the cellular decision-making processes that come into play once the cell's DNA has been damaged. Thus extracellular and intracellular signals, post-translational control mechanisms and the proliferative state of the cells are all crucial factors in the cell's decision to proliferate or to undergo programmed cell death. The volume ends with a provocative look to the future by Nick Hastie. He highlights some of the exciting recent developments in gene regulation and points to the areas where dramatic progress in our understanding can be anticipated, generating much future excitement.

We thank all the contributors for their thought-provoking essays and for addressing the brief we gave them in such exciting ways. We are sure this volume will enable senior undergraduate and junior postgraduate students to appreciate the amazing subtlety and diversity of gene regulatory mechanisms and will hopefully encourage them to participate in the challenging research work that lies ahead in the Third Millennium.

Thanks are also due to our production editor, Sophie Dilley, and her colleagues at Portland Press for the high quality of the volume.

Karen Chapman (Edinburgh)
Steve Higgins (Leeds)
April 2001

Authors

Alexander Gann is Senior Editor at Cold Spring Harbor Laboratory Press. He received his PhD from Edinburgh University in 1988 before pursuing postdoctoral work at Harvard University and the Ludwig Cancer Research Institute at University College London. He was a Lecturer at Lancaster University for two years before moving to Cold Spring Harbor in 1999. **Mark Ptashne** is a professor at Memorial Sloan-Kettering Cancer Center in New York. He gained his PhD from Harvard in 1968. He was a Junior Fellow there before joining the faculty where he remained until moving to Sloan-Kettering in 1998. His research interests are focused on the basic mechanism of gene regulation

Georgina Lloyd obtained her Ph.D. from the University of Leeds, having worked on the purification and characterization of mammalian regulatory aminopeptidases in the laboratory of Tony Turner. Since 1995 she has pursued postdoctoral studies at the University of Birmingham, working with the *Escherichia coli* cAMP receptor protein and concentrating on the role of the RNA polymerase α subunit in transcriptional regulation. **Paolo Landini** obtained his doctorate from the University of Pavia in Italy, studying the interactions between bacterial DNA topoisomerases and DNA. During postdoctoral fellowships at the University of Massachusetts, and in Birmingham, he focused on the *E. coli* Ada protein, which is responsible for the adaptive response to alkylating mutagens. Paolo's work has aimed to understand the interactions of Ada with the different RNA polymerase subunits. Early in 1999, Paolo moved to the Swiss Institute for Environmental Science and Technology in Dübendorf, where he is working on molecular mechanisms of biofilm formation. **Steve Busby's** doctoral studies were concerned with understanding how protein conformation can be controlled by ligands. He was introduced to the marvels of microbial gene regulation and the intricacies of promoters during postdoctoral fellowships at the Institut Pasteur in Paris and the National Institutes of Health in Bethesda, Maryland. In 1983, Steve joined the University of Birmingham and, since then, he has continued active research in the area of microbial gene regulation. He is currently one of the Professors of Biochemistry in the School of Biosciences and is also the Dean of Science.

Grace Gill received her Ph.D. from Harvard University, where she studied the transcriptional activation function of GAL4 with Mark Ptashne. She was a postdoctoral fellow in the laboratory of Robert Tjian at the University of California, Berkeley, where she continued to investigate transcriptional activation mechanisms. Grace Gill is currently an Assistant Professor in the Department of Pathology at Harvard Medical School. She was an Instructor of the Eukaryotic Gene Expression Course at Cold Spring Harbor Laboratory from 1997 to 2000. Current research in her laboratory is directed towards understanding how the activity of specific promoters and transcription factors is regulated during development of the nervous system.

Alan P. Wolffe is Senior Vice President and Chief Scientific Officer of Sangamo BioSciences Inc., a biotechnology company that designs transcription factors for therapeutic applications. From 1990 to 2000, Dr Wolffe was Director of the Department of Molecular Embryology at the National Institutes of Health. Dr Wolffe leads a research group that studies the regulation of gene expression. He has published more than 260 research papers on this topic and is currently an editor of: *Chromosoma, Chemtracts, European Journal of Biochemistry, Gene Therapy and Molecular Biology* and *Molecular Biology of the Cell.* He also serves on the Editorial Boards of: *Biochemistry, Biochemical Journal, Biochimica et Biophysica Acta, BioEssays, Current Genomics, Cell Research, Journal of Cell Science, Molecular and Cellular Biology, Nucleic Acids Research* and *Science.* Dr Wolffe has received several prizes for his research, and he has organized AACR, ASCB and ASMB Symposia, FASEB Conferences, Juan March Foundation Workshops, Keystone Symposia and Novartis Foundation Symposia. He has served on numerous Federal and International scientific advisory boards.

Richard Meehan undertook his Ph.D. at Edinburgh on the genetics of drug detoxification systems in mice, with Nick Hastie and Roland Wolffe. He then worked with Adrian Bird in the Institute of Molecular Pathology, Vienna, where he purified methyl-CpG binding protein 2 (MeCP2), which paved the way for the subsequent isolation of other methyl-CpG binding proteins. He returned to Edinburgh where, at the Institute of Cell and Molecular Biology, where his attempts to isolate MeCP1 revealed it to be a multi-component complex. Subsequently, he became a lecturer at the Department of Biomedical Science at Edinburgh, and began a more rewarding analysis of the role of DNA methylation in *Xenopus* development. **Irina Stancheva** graduated from the University of Sofia, and undertook her Masters degree at the Institute of Cell Biology of the Bulgarian Academy of Sciences, under the supervision of Luchezar Kargyozov. Here, she began working on DNA methylation and developed a technique for formaldehyde cross-linking which allows one to distinguish between active and inactive gene loci. This, together with some studies on DNA re-methylation after the passage of the replication fork, became the topic of her Ph.D., which was awarded by the Swiss Federal Institute of Technology where she worked with Theodor Koller and Jose M. Sogo. She is currently working with Richard Meehan on the role of DNA methylation in *Xenopus* development.

Melanie Lee obtained a B.Sc. in Biochemistry from the University of Manchester Institute of Science and Technology in 1993 and a Ph.D. in Biochemistry from St.George's Hospital Medical School, University of London in 1997. She is currently a post-doctoral fellow at the Marie Curie

Cancer Research Institute where, she is studying the control of gene regulation in melanocytes and melanomas. **Stephen Goodbourn** obtained a B.A. in Biochemistry from the University of Oxford in 1979 and a D.Phil. in Clinical Medicine from the University of Oxford in 1983, and was a post-doctoral fellow in the Department of Biochemistry and Molecular Biology at Harvard University from 1983 to 1987. He was head of the Gene Expression Laboratory at the Imperial Cancer Research Fund in London from 1987 to 1994, and since then has been a senior lecturer in Biochemistry and Immunology at St. George's Hospital Medical School, University of London. His research interests include the control of cytokine gene expression and the interactions between viral infections and cell signalling.

Ho Man Chan is a final year student supported by the Wellcome Trust. **Noriko Shikama** is a postdoctoral fellow, supported by an EMBO Fellowship, having completed her graduate studies at the University of Basel. **Nicholas La Thangue** is the Cathcart Professor of Biochemistry at the University of Glasgow, and was previously a staff scientist at the MRC National Institute for Medical Research.

Chris Proud carried out his Ph.D. work at the University of Dundee on the role of protein phosphorylation in the regulation of glycogen metabolism. He was introduced to the field of mRNA translation during his post-doctoral work in Jenny Pain's laboratory at Sussex University and has remained in this research area ever since. During his subsequent research at Bristol and Kent he explored the regulation of a number of translation factors by phosphorylation and studied the protein kinases and phosphatases which act upon them. He now leads a group at the University of Dundee which applies a wide range of techniques to investigate the structure and regulation of a number of translation factor proteins and their roles in the control of gene expression in mammals and fruit flies.

Martyn Link graduated from the University of Edinburgh in 1997, after which he began an MRC studentship investigating the modulation and mechanism of cell death in the pancreatic β-cell. **David Harrison** is Professor of Pathology at the University of Edinburgh. He is interested in the cell biology of apoptosis and the factors that influence why the same injury may cause different effects in different cell lineages in different environmental conditions. He is a graduate of medicine from Edinburgh University and has clinical interests in the diagnosis of liver and pancreatic disease.

Nick Hastie is Director of the Medical Research Council Human Genetics Unit in Edinburgh. Over his career he has worked in several areas, including developmental gene expression, genome organization, telomeres and developmental genetics. During the past decade he has focused on the childhood cancer Wilms' tumour, and the multiple functions of the Wilms' tumour-suppressor gene, *WT1*. From 1990 to 1997 he was the European Editor of *Genes & Development* and he continues to sit on the Editorial Board of this journal.

Abbreviations

Apaf	apoptotic protease-activating factor
APC	adenomatous polyposis coli
ATF	activating transcription factor
BAF	BRG1-associated factor
BH domain	Bcl-2 homology domain
BRG	*Brahma*-related gene
CAP	catabolite gene activator protein
CBP	CREB-binding protein
CDC	cell division cycle
CDK	cyclin-dependent kinase
CREB	cAMP-response-element-binding protein
CRP	cAMP receptor protein
αCTD	C-terminal domain of RNA polymerase α subunit
DAG	diacylglycerol
Dnmt	DNA methyltransferase
DPE	downstream promoter element
dsRNA	double-stranded RNA
4E-BP	eIF4E-binding protein
eEF	eukaryotic elongation factor
eIF	eukaryotic initiation factor
ER	endoplasmic reticulum
ERK	extracellular signal-related protein kinase
FADD	Fas-associated death domain
GEF	GDP/GTP exchange factor
Grb2	growth-factor-receptor-bound protein 2
GSK	glycogen synthase kinase
GTF	general transcription factor
HDAC	histone deacetylase
HNF	hepatocyte nuclear factor
Inr	initiator (of transcription)
IP_3	inositol 1,4,5-trisphosphate
IRE	iron response element
IRF	interferon regulatory factor
IκB	inhibitor of NK-κB
JAK	Janus kinase
JNK	c-Jun N-terminal kinase
MAPK	mitogen-activated protein kinase
MAPKK	MAPK kinase

MAPKKK	MAPK kinase kinase
MBD	methylated-DNA-binding domain
^{5}mC	cytosine methylated at position 5
MDM2	murine double minute clone 2 oncoprotein
MeCP	methylated-CpG-binding protein
^{Me}CpG	methylated CpG
MEK-1	MAP kinase/ERK kinase
Met-tRNA	methionyl-tRNA
MPTP	mitochondrial permeability transition pore
NFAT	nuclear factor of activated T-cells
NF-κB	nuclear factor κB
αNTD	N-terminal domain of RNA polymerase α subunit
ORF	open reading frame
PDGF	platelet-derived growth factor
PERK	PKR-like ER-resident kinase
PIP$_2$	phosphatidylinositol 4,5-bisphosphate
PIP$_3$	phosphatidylinositol 3,4,5-bisphosphate
PKA	protein kinase A
PKR	dsRNA-activated protein kinase
PLC	phospholipase C
PP	pocket protein
pRb	protein product of the retinoblastoma tumour suppressor gene
PV	polio virus
Rb	retinoblastoma
RNAP	DNA-dependent RNA polymerase holoenzyme
SH	Src homology
SNF	sucrose non-fermenting in Saccharomyces cervisiae
SRB	suppressor of RNA polymerase B
SRF	serum response factor
STAT	signal transduction and activator of transcription
SWI	mating-type switching in Saccharomyces cervisiae
TAF$_{II}$	TATA-box-binding-protein-associated factor
TBP	TATA-box-binding protein
TNF	tumour necrosis factor
TOP	tract of pyrimidine
(m)TOR	(mammalian) target of rapamycin
UPE	upstream element
UTR	untranslated region

Transcription initiation: imposing specificity by localization

Mark Ptashne*[1] and Alexander Gann[†]

*Cold Spring Harbor Laboratory, 1 Bungtown Road, Cold Spring Harbor, NY 11724, U.S.A., and [†]Memorial Sloan-Kettering Cancer Center, 1275 York Avenue, New York, NY 10021, U.S.A.

Introduction

Two broad classes of enzymes may be distinguished by their modes of regulation. Members of the first class, exemplified by the enzymes of intermediary metabolism, recognize one or a few specific substrates, and are regulated by substrate concentration and by allosteric effects exerted by other small molecules. In contrast, members of the second class can recognize a large array of related substrates, the concentrations of which do not vary. This class includes the most common form of RNA polymerase in *Escherichia coli*, protein sorting and degrading enzymes, and the protein kinases and phosphatases of signal transduction pathways.

In this chapter, we will discuss a common and widely used strategy by which enzymes in this second class are regulated. How, for example, one extracellular signal leads to one pattern of gene expression or protein phosphorylation, whereas another directs the same enzymic machinery to produce a different pattern. As a great deal of recent work has revealed, this strategy entails the regulated localization of the enzyme to the appropriate substrate. The term 'localization' means apposition and does not necessarily imply

[1]*To whom correspondence should be addressed
(e-mail m-ptashne@ski.mskcc.org)*

sequestration to particular sites or compartments within the cell. A well understood form of gene regulation presents a particularly well-characterized example of the localization strategy. In this case, locator proteins (transcriptional activators) bring the enzyme (RNA polymerase) to specific promoter sequences. Specificity can be, and typically is, imposed by simple binding interactions between a locator, the transcriptional machinery and the DNA.

Co-operative binding of proteins to DNA

Much of gene regulation depends upon the co-operative binding of proteins to DNA. Co-operative binding is used to direct proteins to specific sites on DNA — that is, to locate them properly — and Figure 1 shows a simple example. As is typical of a DNA-binding protein, the depicted protein recognizes related sequences with different affinities. At its cellular concentration, the protein binds spontaneously to certain sites (strong sites) but leaves others (weak sites) unfilled. However, the protein can be directed to a weak site by interacting with a second protein, which is binding simultaneously at a nearby DNA site. The second protein has located the first to the weak site by increasing the local concentration of the first protein in the vicinity of that site.

Effective use of co-operative binding requires that the concentrations of interacting proteins be controlled. This requirement arises because rather weak interactions (interaction energies of the order of a few kilojoules) between pairs of co-operatively binding proteins usually dictate the reaction. Simply raising the concentration of a protein as little as 10-fold often suffices to pro-

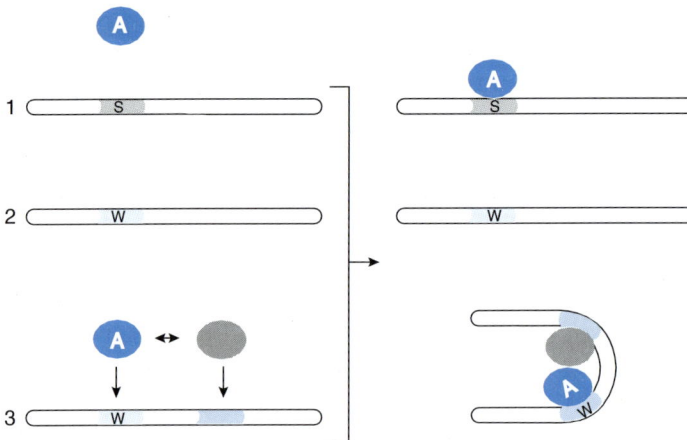

Figure 1. Localization by co-operative binding to DNA
Protein A binds to the strong (S) site on DNA molecule 1, but not to the weak (W) site on molecule 2. However, protein A does bind to the weak site on molecule 3 by virtue of an interaction with, and hence co-operative binding with, another protein that binds to a site nearby.

mote spontaneous binding to weak sites. Consequently, if co-operative binding is the way of regulating localization, the interacting proteins must be maintained below levels at which their interactions become unnecessary for binding.

In the simplest scenario, neither partner of a pair of co-operatively binding proteins needs to undergo a modification or a conformational change. Rather, the interaction between the proteins, as well as that between the proteins and the DNA, need only provide the requisite binding energy. Therefore, these kinds of interactions, which can be and sometimes are highly specific, need only be adhesive (glue- or Velcro-like).

Activators as locators

The most common form of *E. coli* RNA polymerase illustrates several of these general features of DNA-binding proteins (Figure 2). At the intracellular concentration of polymerase, certain promoters constitute strong sites and are therefore recognized spontaneously at high frequency, whereas others are weak and are recognized infrequently. However, genes with either category of promoter can be regulated using the principles described above to produce equally high (or low) levels of transcription and to do so only when appropriate. For example, RNA polymerase can be directed to a specific weak promoter, and the gene thereby activated, by binding co-operatively to the DNA with another protein, called an activator.

The typical activator bears two essential surfaces: one that recognizes a specific site on DNA, and another, the activating region, that interacts with RNA polymerase. Therefore, promoter selection is dictated by specific binding of the activator to a site near one (or another) promoter. As mentioned, in this case, activators would appropriately be called locators. Contrary to the implication of their current name, they literally activate neither the gene nor the polymerase. Rather, they appose specific substrates (genes) with the enzyme, whereupon the elaborate process of transcription proceeds spontaneously.

There are two special cases of activation in *E. coli* that do not conform to this model. In one, a special form of RNA polymerase binds, in an inactive state, to a promoter. In this case, the activator must induce a conformational change in the enzyme in order to activate it. The second example involves a special class of promoter; in this case the activator induces a change in the promoter to activate transcription. These special cases (exemplified by the *glnA* gene in the first case, and the *MerT* gene in the second) provide a useful contrast to the mechanism of localization, and can be distinguished from it experimentally.

One of the crucial experimental strategies that has led to our view of how many transcriptional activators work comprise the so-called 'activator-bypass' experiments. Such experiments show that bacterial and eukaryotic genes acti-

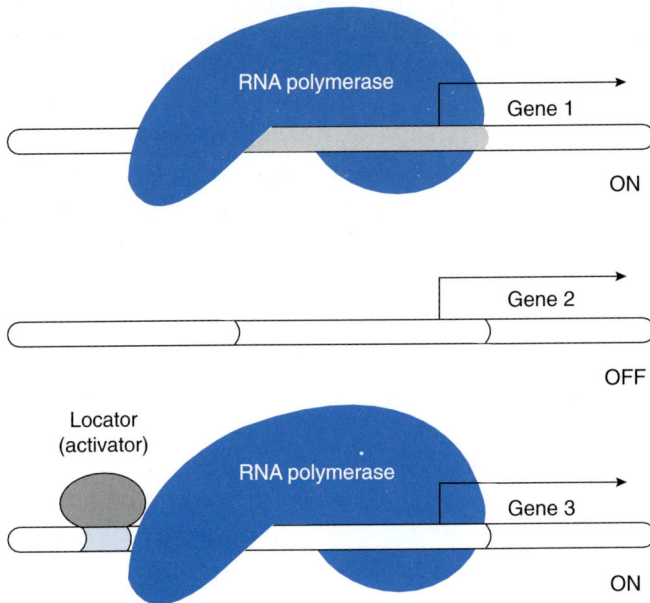

Figure 2. Gene activation as an example of co-operative binding to DNA
The promoter sequence of gene 1 binds polymerase sufficiently tightly that the gene is 'on' in the absence of any activator (and of any repressor that would otherwise prevent polymerase binding). Genes 2 and 3 have weak promoters, and polymerase only binds if helped to do so by an activator (Locator), as illustrated for gene 3.

vated by localization can be activated in the absence of any typical (classical) activator [1]. For example, a DNA-binding domain fused to a component of the transcriptional machinery can activate transcription very efficiently at promoters bearing the appropriate DNA-binding site, in either bacteria or yeast, as can an arbitrary contact between a DNA-tethered peptide and the machinery. Simply increasing the concentration of the bacterial or yeast transcriptional machinery also suffices to mimic the effects of activators *in vitro*. Activator bypass experiments do not work on genes activated by alternative mechanisms.

The simple scheme for gene activation illustrated in Figure 2 readily lends itself to modulation by further co-operative binding. For example, in many instances the activator itself interacts and binds co-operatively with other proteins to DNA. Those additional proteins may or may not be activators themselves, but in either case the result is to make the effect of any activator dependent upon co-operative binding with other proteins. As we shall see, these kinds of auxiliary interactions can be used to make activation of a gene dependent upon more than one physiological signal and to make sensitive switches. There is a further source of co-operativity implicit in this scheme, one that makes it easy to see how activators that do not interact with each other can nevertheless work together synergistically. Any DNA-bound activators that can simultaneously touch the transcriptional machinery would work synergistically because each would contribute binding energy to the recruitment reaction. The observations of unrelated activators working synergistically when placed near a gene are consistent with this expectation and they suggest simple evolutionary pathways for modifying the regulation of genes.

Other proteins, called repressors, prevent access to the promoter and turn off transcription. Repressors themselves can convey extracellular signals, and the combined effects of repressor and activator provide another means for making expression of a gene dependent upon more than one physiological signal.

Allostery — the rest of the story

Extracellular signals that regulate genes are not generally detected by the simple binding interactions of the sort we have been describing. Rather, each such signal is often accompanied by an allosteric change in a target protein. For example, upon binding to a metabolic derivative of lactose, the Lac repressor of *E. coli* undergoes a structural transition that prevents it from binding DNA [2]. This and many other examples suggest a generalization: allosteric-like interactions are typically used to reveal the presence of an extracellular signal, but the specific interpretation of that signal is then dictated by the localization mechanisms. Moreover, the meaning of any given signal can be changed or expanded without changing the allosteric response itself. For example, the Lac repressor detects the presence of lactose, but that condition can be used to repress any gene depending upon the disposition of the repressor binding sites on DNA.

Examples of gene regulation in bacteria

We shall consider the action of two well-studied bacterial transcriptional activators: catabolite gene activator protein (CAP) and lambda repressor (which despite its name can function as an activator – see later). The ability of each to bind DNA, and hence to function, is determined by extracellular signals that induce changes in the proteins. CAP functions only in the absence of glucose (see Chapter 2 in this volume by Lloyd, Landini & Busby), and lambda repressor is inactivated when DNA is damaged by agents such as ultraviolet light [3]. The specificity of action of each protein (which gene it regulates) is determined by its DNA-binding address. CAP ordinarily binds to, and thus activates, sites near genes encoding enzymes required for metabolism of various sugars, while lambda repressor ordinarily activates its own gene [3]. If a CAP site is introduced upstream of the promoter of the lambda repressor gene, CAP will activate that gene in the absence of glucose [4]. The meaning of the physiological signal, in this case the absence of glucose, can thus be re-interpreted simply by introducing the relevant DNA site in front of a gene.

The activities of CAP and lambda repressor illustrate two additional features expected of locators that work as outlined above. Firstly, if both CAP and lambda repressor are positioned adjacent to a promoter so that each can make its natural contact with RNA polymerase, the two activators work synergistically, as expected if the proteins simultaneously contact polymerase and work together to recruit polymerase [4]. Secondly, each protein can, and at certain promoters does (as the name lambda repressor suggests), work as a repressor. All that is required is that the protein be positioned so that, rather than making a fruitful contact with polymerase, it blocks polymerase binding [3,5].

Multiple signals and combinatorial control

Control of sugar metabolism genes in *E. coli* illustrates how gene expression can be made dependent on two signals by regulators that do no more than help or hinder polymerase localization. The example also illustrates how regulators can be used in different combinations. *E. coli* bears separate sets of genes, each of which encodes enzymes that direct metabolism of one or another of a wide array of sugars. The biological problem is to ensure that any set of such genes is expressed only if two conditions hold: (i) that the relevant sugar, such as lactose or galactose, is present in the medium and (ii) that glucose, a better carbon source, is absent.

Figure 3 shows how this is achieved for the *lac* genes. These genes are activated by CAP (which, as we have noted, is only active in the ~~presence~~ *absence* of glu-

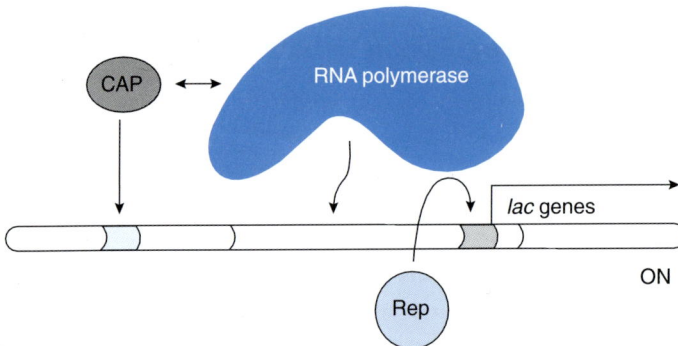

Figure 3. Control of gene expression by the *lac* promoter in *E. coli*
In the absence of any controlling factors, and at intracellular concentrations, polymerase transcribes the genes at a low level. Transcription is increased some 50-fold by CAP, which binds just upstream of the polymerase and, by simultaneously contacting polymerase with its activating region, binds co-operatively with it. The Lac repressor (Rep) has the opposite effect. It binds to a site in the promoter that overlaps sequences which would otherwise be contacted by RNA polymerase and thereby prevents transcription. CAP and Lac repressor respond to separate physiological signals allosterically. CAP binds DNA only when complexed with cAMP, which is depleted by growth in glucose. Lac repressor cannot bind DNA when complexed with a metabolite of lactose.

cose) provided that lactose, which inactivates the Lac repressor, is also present. Regulators that work as described in Figure 3 readily lend themselves to being used in different combinations. For example, a CAP site is also located upstream of the *gal* genes, where CAP works with the Gal repressor to control transcription. Thus CAP activates the *gal* genes in the absence of glucose, provided that galactose is simultaneously present to inactivate the Gal repressor. CAP works in combination with many other regulators at some 100 genes in *E. coli* (see Chapter 2 in this volume by Lloyd, Landini & Busby).

It is not difficult to imagine how systems such as this evolved from a rudimentary system that worked but was inefficient. For example, in the absence of binding sites for the regulatory proteins, the weak *lac* promoter would read at a constant and low level. The bacterium would be able to use lactose, but it would make the enzymes even when there was no lactose substrate and also when the superior carbon source, glucose, was present. The first improvement would be the addition of a CAP-binding site, positioned so that CAP would contact polymerase at the promoter and hence bind co-operatively to DNA with it. The system would now provide high levels of the enzyme in the absence of glucose and lower levels in its presence, irrespective of the presence of lactose. A further refinement would be the addition of a binding site for Lac repressor, which would ensure that transcription is switched off in the absence of lactose.

Using simple binding interactions to make a sensitive switch

Phage lambda shows how simple binding interactions can create a switch that responds in an all-or-nothing fashion to an extracellular signal (Figure 4). The biological problem is that within an *E. coli* host cell, the genes of the bacterial virus lambda can be maintained in a silent state, known as lysogeny, until an inducing signal is detected. They then are efficiently activated, leading to lytic growth [3]. This regulatory problem has been solved by constructing a biphasic switch involving two adjacent promoters. These promoters are controlled according to the rule that when one is on the other is off. Here we find two forms of co-operativity in addition to that involving an activator and RNA polymerase. These additional features are crucial to the efficiency of the switch.

The key regulator is the lambda repressor, a protein that simultaneously activates transcription of its own gene as it turns off other genes. As shown in Figure 4, two DNA-bound repressor dimers are positioned so that they cover and turn off the strong rightwards promoter, P_R, which controls the lytic genes. Simultaneously, one of these repressors contacts RNA polymerase and activates transcription of the weak leftwards promoter, P_{RM}. This activation ensures that, once repressor synthesis has been initiated (an event that requires a separate promoter and activator), the repressor maintains its own synthesis. The phage genome is thereby stably maintained in a near-silent state, the only active gene being that of the repressor itself. The system stably perpetuates itself until the cell encounters the signal that triggers the switch mechanism.

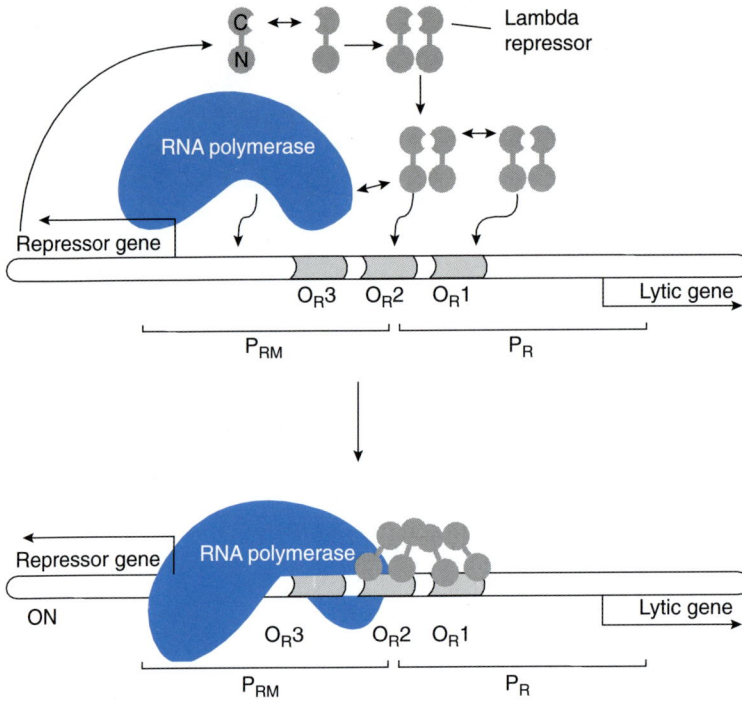

Figure 4. The phage lambda switch
Repressor monomers comprising two domains (C and N) separated by a linker are in equilibrium with dimers, the DNA-binding species. Two repressor dimers bind co-operatively to the adjacent operator sites O_R1 and O_R2. Repressor at these two sites represses the lytic promoter P_R, a strong promoter that works spontaneously at a high level unless repressed. Simultaneously, repressor activates the weak promoter of the repressor gene itself, P_{RM}, by virtue of a contact between repressor at O_R2 and polymerase at P_{RM}. At higher concentrations, repressor also binds to O_R3 and turns off P_{RM}, thereby negatively regulating repressor synthesis. The three surfaces on the repressor involved in the three examples of co-operativity (repressor dimerization, interactions between dimers, and interaction with polymerase to activate P_{RM}) are shown in white. As described in the text, repressor is cleaved in response to ultraviolet radiation, and as a consequence transcription from P_R is turned on as that from P_{RM} is turned off.

Then, as repressor is inactivated, the rate of further repressor synthesis also drops. The first gene transcribed upon induction, *cro*, encodes another repressor that turns off P_{RM}, thus further ensuring that induction of lytic growth is an all-or-nothing effect.

The two additional forms of co-operativity in the lambda switch mediate co-operative binding of the repressor to DNA. In the cell, repressor monomers are in concentration-dependent equilibrium with dimers, the DNA-binding species. Two repressor dimers bind co-operatively to the adjacent operator sites, as shown in Figure 4. These repressor–repressor interactions ensure that the operator sites are filled as a highly sigmoidal function of the repressor concentration, providing both a buffer against minor fluctuations in repressor concentration and a dramatic change in state when some significant, but readily obtainable, proportion of repressor (approximately 90%) is inactivated.

A remarkable feature of the switch is that it depends upon a series of weak protein–protein interactions. Under physiological conditions P_{RM} is only activated by a factor of about 10 and co-operative binding to the two adjacent sites also has just a 10-fold effect. Each of these interactions therefore requires only 4–9 kJ of binding energy, an amount easily provided by a simple protein–protein interaction. The requirement for each of the three protein–protein interactions can be dispensed with simply by increasing the concentration of one of the components. For example, increasing the concentration of polymerase *in vitro* is sufficient to elicit activated levels of transcription from P_{RM} in the absence of repressor [6]. In addition, although binding of repressor to the site adjacent to the polymerase (O_R2) ordinarily depends upon interaction with another repressor dimer binding to the auxiliary site (O_R1), merely increasing the repressor concentration some 10-fold obviates the need for this latter interaction. Repressor then binds spontaneously to O_R2 and performs both of the required functions (activation and repression). Thus, although repressor at O_R1 also helps repress P_R, its uniquely required function is to impose co-operativity on the system.

As illustrated in Figure 4, the three protein–protein interactions seen in the switch (repressor dimerization, co-operative binding of repressor dimers, and interaction with polymerase) involve separate patches on the surface of the repressor. Nevertheless, it is likely that, as expected for a series of simple binding interactions, they are interchangeable. For example, the protein–protein interaction between repressor and polymerase responsible for activation can be replaced by the one that normally mediates co-operative binding. This is an example of the activator-bypass experiment described above. In this case, polymerase is modified so as to bear a pair of lambda repressor C-terminal domains. Interaction of these domains with those of a lambda repressor bound to DNA nearby suffices for gene activation. The interaction that ordinarily mediates co-operative binding of lambda repressors can thus equally well mediate transcriptional activation [7].

The importance of repressor dimerization and the interaction of repressor dimers in promoting co-operative DNA binding is demonstrated by the fact that induction by ultraviolet light works simply by eliminating these functions. The induction process is mediated by a protein, RecA, which recognizes DNA and undergoes an allosteric transition that activates its protease function. Repressor is cleaved at a specific peptide sequence linking the two domains, N- and C-terminal, of the protein. The N-terminal domain is still capable of carrying out the essential functions of the intact repressor — DNA binding (and hence repression of one set of genes) and contact with polymerase (and hence activation of the repressor gene) — but at intracellular concentrations it fails to do so in the absence of the co-operative effects mediated by the C-terminal domain. The sole function of the C-terminal domain is to promote dimer formation and interaction between dimers. Separating the two domains eliminates both forms of co-operativity, and is sufficient to trigger induction.

Herein lies the problem incurred by relying upon relatively weak binding interactions to impose specificity by localization. The concentrations of the components must be maintained over a relatively narrow range. This is accomplished here by the imposition of a third repressor binding site, O_R3, that overlaps P_{RM}. Repressor bound to O_R3 blocks polymerase binding to P_{RM} and thus negatively regulates its own synthesis. Repressor binds co-operatively to O_R1 and O_R2 with an affinity some 10-fold higher than that with which it binds to O_R3. O_R3 therefore only becomes relevant at higher repressor concentrations. This simple governing mechanism ensures that repressor never reaches a concentration at which it can bind to O_R2 without binding co-operatively with another repressor dimer binding to O_R1.

As might be expected from this line of analysis, genes encoding many transcriptional regulators (including those encoding the subunits of RNA polymerase) are regulated so as to ensure that the concentrations of their products are maintained below specified levels [5,8].

Gene regulation in eukaryotes

We noted above an experiment in which a bacterial gene was brought under control of a heterologous activator (CAP) merely by introducing the binding site for that regulator near the gene. Similar experiments have been performed with many activators and genes in many eukaryotes. The experiment is actually easier to perform in eukaryotes as a successful outcome is much less dependent upon precise positioning of the activator relative to the gene. Two factors evidently contribute to this greater flexibility. Firstly, a typical eukaryotic activator apparently binds more tightly to its targets in the transcriptional machinery than does a typical bacterial activator and hence will work from further upstream. Secondly, a typical eukaryotic activating region can evidently contact several, perhaps many, sites on the transcriptional machinery. The latter property may be particularly important in allowing the activator to work at a wide array of promoters. Depending on the position of the activator on the DNA in relation to the transcriptional start site, certain contacts may be used in place of others.

Eukaryotes have widely exploited combinatorial strategies to create gene regulatory networks. Many eukaryotic genes, especially in higher organisms, respond in a switch-like fashion to multiple signals. That is, the gene is on if, and only if, several physiological signals are detected simultaneously. The following example shows how these mechanisms are used to create such a switch for the human interferon-β gene (Figure 5). Here, three separate activators, nuclear factor κB (NF-κB), activating transcription factor (ATF)/Jun and interferon regulatory factor (IRF)3/7, bind DNA co-operatively to form an

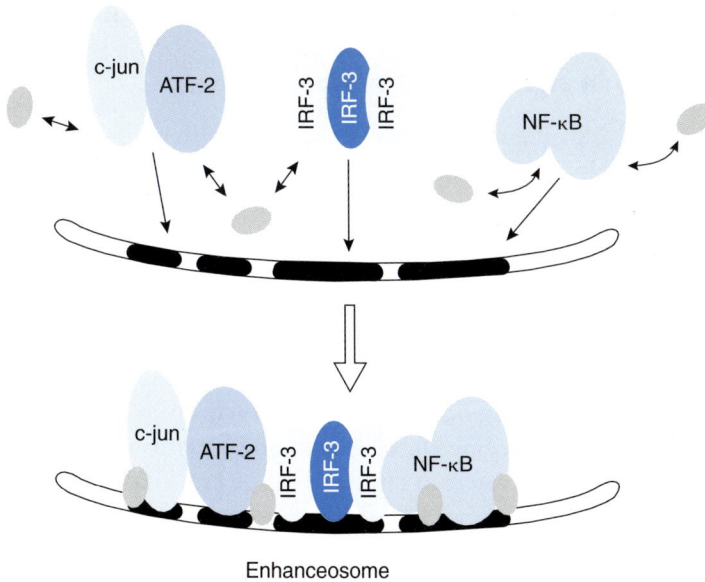

Enhanceosome

Figure 5. Human interferon-β enhanceosome

Three transcriptional activators (NF-κB, ATF/Jun and IRF-3/7) are activated in response to virus infection. The mechanism of activation is different for each transcription factor. For example, NF-κB is released from a bound inhibitor and allowed to enter the nucleus, and the DNA-binding function of ATF/Jun is activated by phosphorylation. These transcriptional activators, interacting with each other and with the auxiliary protein HMG-γ (light grey), bind co-operatively to DNA to form the enhanceosome.

enhanceosome. Because of the co-operativity, this requires that each of the activators receives its appropriate physiological signal, rendering it capable of binding to DNA. Virus infection, which produces all three signals, thus triggers formation of the enhanceosome and activates the interferon-β gene [9,10]. Once the enhanceosome has formed, the activating regions carried on its various constituents simultaneously contact the transcriptional machinery and thereby work synergistically to activate transcription.

Optimal functioning of the interferon-β enhanceosome requires rather precise spacings between the binding sites for the components listed above, as well as for certain auxiliary proteins. Those spacings ensure that the various components can simultaneously touch one another, DNA and the transcriptional machinery. The precise positioning of the enhanceosome with respect to the promoter is not critical, however, and the enhanceosome functions when positioned at any of the many sites within hundreds of base pairs of the gene.

Enzymes that modify nucleosomes can play an important role in modifying gene expression (see Chapter 4 in this volume by Wolffe). These enzymes are also regulated by localization. Their specificity of action, that is which nucleosomes they modify, is determined by interaction with specific DNA-binding transcriptional activators and repressors. In the absence of such specificity determinants, these enzymes work on all nucleosomes at a lower level.

Localization in signal transduction

We noted in the Introduction that many biological systems use the principle of imposing specificity by localization. Two examples from signal transduction systems will be used to illustrate aspects of localization encountered in our discussion of gene regulation.

Signal transduction and activators of transcription (STATs)

We suggest that there is an analogy between the workings of a receptor, here a cytokine receptor, and transcriptional activators like CAP. As we have seen, the latter detect external signals and interpret them by working as locators, bringing together an enzyme (RNA polymerase) with one or other of its potential substrates (promoters of target genes). A cytokine receptor responds to a signal by bringing together an enzyme (a kinase) with one or other of its

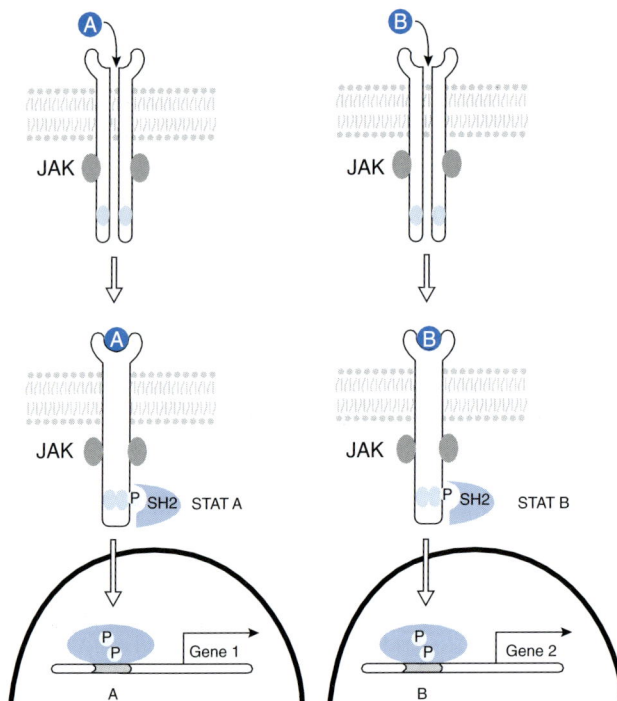

Figure 6. STAT activation
Cytokine A activates gene 1 by inducing phosphorylation (P) of STAT A, whereas cytokine B activates gene 2 by inducing phosphorylation of STAT B. If the STAT A binding site on the cytokine A receptor is replaced by a site that binds STAT B, cytokine A activates gene 2 instead of gene 1. JAK (Janus kinase).

potential substrates (so-called STAT proteins). The specificity of the response (i.e. which STAT is phosphorylated) is determined by simple binding interactions and is therefore readily changed [11]. The cytokine system (described in detail in Chapter 6 in this volume by Lee & Goodbourn) is illustrated in Figure 6. Two different cytokines interacting with their respective receptors activate two different STATs. Receptor recognition dimerizes the receptor and triggers phosphorylation of receptor tyrosine residues [12,13]. Phosphorylation creates a specific STAT-binding site. The bound (localized) STAT is positioned close to JAK (a Janus kinase) which phosphorylates and activates it. The activated STAT, now a dimer, moves to the nucleus and activates specific genes.

The identity of the STAT activated by a given cytokine is determined by which STAT binds the receptor. That specificity is readily altered. Interchanging STAT-binding sites between receptors or receptor-binding sites [the so-called Src homology 2 (SH2) domains] between STATs, suffices to swap the specificities of the responses. Thus the specificity of the response to a given cytokine is determined not by the inherent specificity of the kinase, but by the identity of the STAT that is brought into the vicinity of the specific receptor and its associated kinase.

Ras

An important aspect of the localization idea is that once the enzyme (RNA polymerase) is brought to the substrate (promoter) the enzymic activity (transcription) proceeds spontaneously. Experiments in which RNA polymerase was brought artificially to the gene have been crucial in formulating our ideas. The same experimental approach has revealed the sufficiency of localization in another signal transduction pathway, that involving the small GTPase, Ras.

Many receptor tyrosine kinases, such as the epidermal growth factor receptor, exert their effects through the Ras pathway, a series of interactions between components widely conserved in eukaryotic evolution. Once again, phosphorylation of sites on the receptor in response to the extracellular signal creates a binding site for, and thus recruits, another protein, in this case Grb2. The adaptor protein Grb2 in turn binds and recruits to the membrane another protein, Sos. This then interacts with and activates a membrane-bound Ras by promoting exchange of GTP for bound GDP. Ras in turn recruits and activates Raf, a kinase that initiates the so-called mitogen-activated protein kinase (MAPK) cascade. This results finally in activation of various proteins, including a number of transcription factors (see Chapter 6 in this volume by Lee & Goodbourn).

In an experiment analogous to an activator-bypass experiment, Sos (actually a fragment of Sos lacking an autoinhibitory domain) was artificially tethered to the membrane by myristoylation and Ras was found to be activated as a result [14]. Thus an important, and perhaps the sole, role of the upstream components of this pathway is to recruit Sos to the membrane in response to the appropriate signal, where it can work on Ras. As would be predicted from this result, over-production of a fragment of Sos without specific recruitment to the membrane also activates the Ras pathway, albeit weakly [15].

These examples (STATs and Ras) show how simple binding interactions localize enzymes and give specificity to signals. In this regard protein–protein interaction motifs (e.g. SH2, SH3, etc.) play the same role as the motifs that direct regulatory proteins to specific sites on DNA. Just as for transcriptional regulatory systems, these signalling pathways are particularly 'evolvable'. For example, it is easy to see how the meaning that a cell ascribes to a particular cytokine can be changed or expanded by attaching binding sites for the appropriate STAT to its receptor. New responses can thus be generated without the need to evolve new enzymic activities or specificities, a requirement that would presumably be more taxing. A similar theme is encountered with the MAPK (see Chapter 6 in this volume Lee & Goodbourn).

An alternative world

Why is the strategy of imposing specificity by localization found so widely in Nature? Consider, for example, control of transcription. One could imagine a system in which specificity is determined purely by allosteric control. In such a system, there would be many different RNA polymerases, activation of any one only being triggered upon integration of the required signals that would together induce an allosteric transition in the appropriate polymerase. Such a system might appear more simple, in some regards at least, than that which is observed. For example, there would be no need for locators or the elaborate use of co-operativity of the type we have described.

The first difficulty in constructing such a purely allosteric world would be to design polymerases that would each integrate the effects of multiple signals. For example, at the *lac* promoter the polymerase would have to be active only when lactose was present and glucose absent. The problem would be magnified in higher eukaryotes where, as we have seen, the presence or absence of multiple signals is often integrated in the decision as to whether a given gene is transcribed. Even if these design problems were solvable [16], it seems likely that it would be more difficult to use the principle of combinatorial control in designing new polymerases that responded to new combinations of signals. That is, whereas locators can readily be used combinatorially, it is difficult to imagine that allosteric modules (if they existed) could be used so flexibly.

Summary

- *Many crucial cellular enzymes – including RNA polymerases, kinases, phosphatases, proteases, acetylaters, etc. – have multiple potential substrates. Regulation entails substrate selection, a process effected by a mechanism we call regulated localization.*

- *This formulation is particularly well illustrated by the mechanisms of gene regulation.*
- *Analysis of these mechanisms reveals that regulated localization requires simple molecular interactions.*
- *These molecular interactions readily lend themselves to combinatorial control.*
- *This system of regulation is highly 'evolvable'.*
- *Its use accounts, at least in part, for the nature of many of the complexities observed in biological systems.*

This essay is a modified version of an article first published in *Current Biology* (**8**, R812–R822, 1998). The authors thank their many colleagues who kindly commented on the original manuscript, and Renate Helmiss for preparing the original illustrations. Figures 1–6 are all reprinted from *Current Biology*, **8**, M. Ptashne and A. Gann, Imposing specificity by localization: mechanism and evolvability, R812–R822, © (1998), with permission from Elsevier Science.

References

1. Ptashne, M. & Gann, A. (1997) *Nature (London)* **386**, 569–577
2. Muller-Hill, B. (1996) *The lac Operon: a Short History of a Genetic Paradigm*, Walter De Gruyter
3. Ptashne, M. (1998) *A Genetic Switch, Phage Lambda and Higher Organisms, 2nd edn*, original printing 1992, Cell and Blackwell Scientific, Cambridge, MA
4. Joung, J.K., Koepp, D.M. & Hochschild, A. (1994) *Science* **265**, 1863–1866
5. Busby, S. & Kolb, A. (1996) The CAP modulon in *Regulation of Gene Expression in Escherichia coli* (Lin, E.C.C., ed.), pp. 255–279, RG Landes, Georgetown, TX
6. Meyer, B.J. & Ptaschne, M. (1980) *J. Mol. Biol.* **139**, 195–205
7. Hochschild, A. & Dove, S.L. (1998) *Cell* **92**, 597–600
8. Dykxhoom, D.M., St Pierre, R., Van Ham, O. & Linn, T. (1997) *Nucleic Acids Res.* **25**, 4209–4218
9. Wathelet, M.G., Lin, C.H., Parekh, B., Ronco, L.V., Howley, P.M. & Maniatis, T. (1998) *Mol. Cell* **1**, 507–518
10. Carey, M. (1998) *Cell* **92**, 5–8
11. Darnell, Jr, J.E. (1997) *Science* **277**, 1630–1635
12. Austin, A.J., Crabtree, G.R. & Schreiber, S.L. (1994) *Chem. Biol.* **1**, 131–136
13. Weiss, A. & Schlessinger, J. (1998) *Cell* **94**, 277–280
14. Aronheim, A., Engleberg, D., Li, N., Al-Alawi, N. & Schlessinger, J. (1994) *Cell* **78**, 949–961
15. Wang, W., Fisher, E.M.C., Jia, Q., Dunn, J.M., Porfiri, E., Downward, J. & Egan, S.E. (1995) *Nat. Genet.* **10**, 294–300
16. Liu, X., Guy, H.I. & Evans, D.R. (1994) *J. Biol. Chem.* **269**, 27747–27755

2

Activation and repression of transcription initiation in bacteria

Georgina Lloyd, Paolo Landini[1] and Steve Busby[2]

School of Biosciences, The University of Birmingham, Edgbaston, Birmingham B15 2TT, U.K.

Introduction

A remarkable feature of bacteria is their ability to respond to environmental stimuli and to adapt to changing environmental conditions. Rapid adaptation is achieved by switching on and off the expression of specific genes. Bacteria are very efficient at increasing the production of certain proteins and enzymes when they are needed, only to shut off their production when it becomes metabolically and energetically wasteful. Tight control of bacterial gene expression can be achieved at many levels, including regulation of transcription initiation, transcript elongation, translation, messenger RNA stability or availability and protein turnover. These mechanisms are not mutually exclusive, and expression of some genes is subject to all the above levels of regulation. In this chapter we focus on regulation at the level of transcription initiation simply because, in most cases, it is the dominant form of regulation. Most of this chapter concerns regulation in the simple enteric bacterium *Escherichia coli*, which is rightly regarded as a paradigm that can instruct our

[1]*Present address: Department of Microbiology, Swiss Federal Institute for Environmental Science and Technology, 133 Uberlandstrasse, CH-8600 Dubendorf, Switzerland.*
[2]*To whom correspondence should be addressed (e-mail: s.j.w.busby@ bham.ac.uk).*

understanding of other bacterial systems. Additionally, *E. coli* was the first organism in which the molecular biology of gene expression was investigated in depth. The logic of these early investigations has been repeatedly applied to other more complex systems. An excellent account of these early studies and their relevance today can be found in [1].

Transcription regulation

The complete genome sequence of *E. coli* has revealed that the number of potential genes is 4290 [2]. However, the protein responsible for all transcription, DNA-dependent RNA polymerase holoenzyme (RNAP), is present at only 2000 molecules/cell (see [3] for an account of the quantitative aspects of transcription). Thus, because of the relative scarcity of the transcription apparatus, *E. coli* cells need to be very efficient in directing transcription in response to stimuli and stresses and, because of this, gene regulation in bacteria is principally mediated at the level of transcription initiation. This regulation is an elaborate process, involving many proteins. The most important of these proteins are the RNAP σ subunits, which confer different promoter specificity and therefore direct RNAP towards different sets of genes. Next, there is a set of ~150 transcription regulators that up-regulate and down-regulate expression from specific sets of promoters in response to particular environmental signals. Control of bacterial gene expression at the level of transcription initiation has been described for genes involved in almost all cellular processes, from energy metabolism to DNA repair and replication, and to responses to environmental stresses. At some promoters, more than one transcription factor can interact and co-regulate transcription. Thus, we first give a brief outline of the bacterial RNAP, placing particular emphasis on the σ subunit. Next, we describe how transcription factors interact with RNAP to up-regulate and down-regulate expression from promoters. Finally, we consider how promoters are designed to respond to multiple signals transduced via different transcription regulators.

Promoter recognition by RNAP — role of the σ subunit

E. coli RNAP consists of five subunits: two identical α subunits and one copy each of the β, β' and σ subunits [4]. RNAP ($\alpha_2\beta\beta'\sigma$) can be separated into two components: the core enzyme ($\alpha_2\beta\beta'$) and the σ subunit. The core enzyme can bind to DNA in a non-specific fashion and retains the catalytic activity of RNAP, i.e. the ability to synthesize RNA from a DNA template. However, the core enzyme cannot initiate transcription at promoters; the σ subunit is necessary for promoter recognition and regulated transcription initiation [5]. Assembly of a σ subunit into RNAP introduces a major change in the RNAP–DNA interaction, reducing non-specific binding to DNA and conferring the ability to recognize particular promoters. This σ-subunit-dependent RNAP–promoter interaction involves several steps. Initial binding to the promoter results in the so-called 'closed complex', which is then converted into the 'open complex' by local denaturation ('melting'), of the DNA sequence immediately surrounding the transcription start point. On formation of the open complex, the synthesis of the transcript is initiated. RNAP then moves away from the promoter, and after about 20 nt have been incorporated into the nascent RNA chain, the σ subunit is released.

If the core enzyme is considered as the 'constant' part of RNAP, the σ subunit is the 'variable' part. The *E. coli* genome encodes seven different σ subunits; incorporation of each different σ subunit into RNAP results in recognition of different sets of promoters [6]. Most genes expressed in rapidly growing cells are transcribed from promoters dependent on the σ^{70} subunit, which is considered the main σ subunit in *E. coli*. The σ^{70} protein contains 613 amino acids and is the product of the *rpoD* gene. Despite being responsible for the specific binding of RNAP to DNA, σ^{70} does not bind to DNA as a free subunit, but only when assembled into RNAP. This is due to the inhibitory activity of the σ^{70} N-terminal region (region 1). Deletion of this region allows σ^{70} to bind to DNA as a free subunit. A typical σ^{70}-dependent promoter contains two conserved hexamer sequences located near positions −10 and −35 relative to the transcription start point. The consensus sequences for these two promoter elements are TATAAT and TTGACA, respectively. These elements are recognized by distinct regions of the σ^{70} subunit: region 2 for the −10 sequence and region 4 for the −35 sequence (Figure 1).

Alternative σ subunits are involved in the induction of specific sets of genes that are required for adaptation and cell survival during environmental stress [5–7]. For example, σ^{32} is required when *E. coli* is grown at temperatures higher than the optimal 37°C, and triggers the expression of the heat-shock

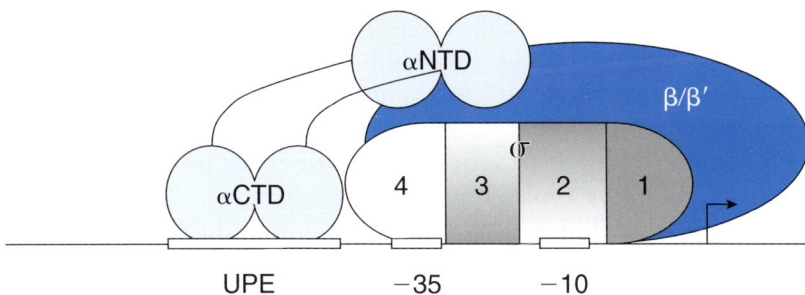

Figure 1. Interactions between RNAP and promoter elements at an activator-independent promoter

The σ subunit of RNAP is responsible for contacting the −10 and −35 hexamers and is shown divided into four distinct regions (shaded). Region 2 and region 4 are involved in promoter recognition; they contact the −10 sequence and −35 sequence, respectively. The α subunits of RNAP consist of two domains, αNTD and αCTD, that are joined by a flexible linker. αCTD provides additional sequence-specific protein–DNA interactions by binding the upstream element (UPE), usually located immediately upstream of the −35 hexamer, thus strengthening RNAP binding. An arrow indicates the transcription start point and the direction of transcription.

genes. A further σ subunit, σ^E, is responsible for transcription of the so-called extreme heat-shock genes. Another example concerns the transition from exponential growth to stationary phase, when a set of genes essential for survival is expressed. The σ subunit that controls expression of these genes is σ^S. In addition to controlling many stationary-phase-specific genes, σ^S is involved in several stress responses. It controls the expression of some genes involved in oxidative stress and in the adaptive response to methylation damage even during exponential bacterial growth.

Subunits of the RNAP core

The structure at 3.3 Å resolution of core RNAP from *Thermus aquaticus* has been reported [8]. Sequence analysis suggests that this structure, although not identical with that of the *E. coli* core RNAP, is very similar. The structure shows a large cleft, which accommodates the template DNA. One face of the cleft is lined by the β subunit and the opposite face is lined by the β′ subunit. Side chains from the β and β′ subunits form the enzyme active site and the binding sites for template DNA and product RNA.

The β subunit (151 kDa; encoded by the *rpoB* gene) can be considered as the main catalytic subunit of RNAP. It binds ribonucleoside triphosphates and promotes polymerization of the RNA chain. The β subunit is the target of the transcription inhibitors, rifampicin and streptolydigin. It also carries the target for the alarmone molecules, ppGpp and pppGpp, that are synthesized as part of the stringent response triggered by starvation for amino acids. Alarmone binding to the β subunit shuts down transcription of ribosomal RNA by preventing transcription initiation.

The function of the β′ subunit (155 kDa; encoded by the *rpoC* gene) is not fully understood. The β′ subunit is rich in positively charged amino acids, and is thought to be the main contributor to non-specific binding by core enzyme to the DNA template. Indeed, the β′ subunit binds avidly to DNA and to heparin, a polyanion whose structure mimics DNA. The β′ subunit has been found to be the target of the N4 bacteriophage single-stranded DNA-binding protein during transcriptional activation of the N4 late promoters. This suggests that β′ also plays a role in transcription initiation, either directly or via interaction with the σ subunit.

The RNAP α subunit (37 kDa; encoded by the *rpoA* gene) is the only subunit that is present as a dimer. It performs three critical functions [9] — it is the initiator of RNAP assembly, it contributes to promoter recognition by sequence-specific interaction with DNA and it interacts with many transcription factors, both in transcription initiation and in anti-termination. Structural and biochemical studies indicate that α consists of two independently folded domains with distinct functions: the N-terminal domain (αNTD) of roughly 230 amino acids and the C-terminal domain (αCTD) of approximately 80 amino acids, linked by a flexible linker region (Figure 1). This α linker region is highly sensitive to protease cleavage. αNTD carries the determinants for α dimerization, the first step in RNAP assembly, which follows the pathway $2\alpha \rightarrow \alpha_2 \rightarrow \alpha_2\beta \rightarrow \alpha_2\beta\beta' \rightarrow \alpha_2\beta\beta'\sigma$. αCTD carries the determinants for recognition of a 20 bp promoter element known as the upstream element (UPE), first identified in the promoters of the genes encoding ribosomal RNA [10]. At these promoters, the UPE, which is located just upstream of the -35 and -10 elements, is a major contributor to promoter activity, stimulating transcription initiation by at least 30-fold. The binding of αCTD to the UPE strengthens RNAP binding (Figure 1). In addition, αCTD carries the contact sites for a large number of transcription activators, including the well-characterized cAMP receptor protein (CRP) [11]. Different transcription activators target different regions of αCTD, but in most cases, the interaction appears to stimulate the binding of αCTD, and hence the rest of RNAP, to the promoter DNA (Figure 2a). Interestingly, at some promoters, CRP also interacts with the αNTD, showing that αNTD can also function as a target of transcription activators [12]. Note that some of the determinants in αCTD that are important for interaction with the UPE and with transcription activators are also involved in the interaction with termination and anti-termination proteins, and possibly bind to the nascent RNA chain. This suggests parallels between transcription initiation and the resumption of transcription elongation during anti-termination.

Negative and positive control of transcription initiation

The expression of many bacterial genes is regulated at the initiation of transcription by transcription factors that interact at or near the corresponding promoter. The majority of transcription factors are DNA-binding proteins, and their promoter specificity is determined by preferential binding to particular base sequences. Most transcription factors function as dimers, or as higher multimers. In many cases, each subunit of the dimer (or multimer) contributes to specific DNA binding. Regulation results from the interaction of such transcription factors with promoter DNA and often also involves interactions with RNAP. These transcription factors can increase the rate of transcription initiation (activators) or prevent RNAP from initiating transcription (repressors). As mentioned above, during transcription initiation RNAP progresses through the 'closed complex' into the 'open complex', in which the promoter DNA is unwound at the transcription start. At this stage RNAP initiates RNA synthesis and must clear the promoter before entering the elongation phase of transcription. Activators and repressors can, at least in theory, affect all these steps: initial binding of RNAP to the promoter, open-complex formation, or promoter clearance [13]. A surprising discovery was that some transcription factors can function as both activators and repressors. Examples of such proteins are the bacteriophage λ repressor, which inhibits transcription from the P_R and P_L promoters but activates transcription at P_{RM},

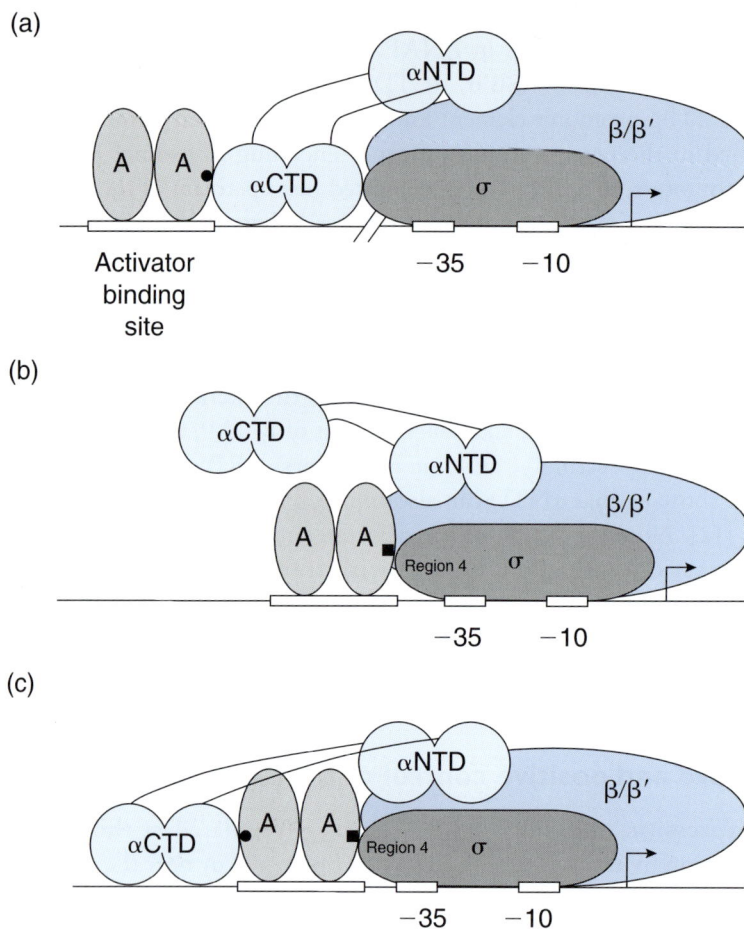

Figure 2. Activation of transcription by activator–RNAP interactions
In each diagram, the activator is shown as a dimer (e.g. as for CRP, bacteriophage λ cl protein, etc.). (a) The activator (A) makes a specific protein–protein interaction (●) with αCTD, which recruits αCTD to the DNA and increases the overall affinity of RNAP for the promoter. (b) The activator (A) makes a specific protein–protein interaction (■) with region 4 of σ, to either recruit RNAP to the promoter or enhance a subsequent step in transcription initiation. αCTD is shown displaced and making no specific contacts with the promoter DNA. (c) The activator (A) contacts both αCTD and region 4 of σ (● and ■, respectively). Each interaction contributes to transcription initiation.

and the CRP protein, which activates transcription of many *E. coli* genes (the *lac* operon is the best known example) but represses others (e.g. the *pro*P1 promoter). It appears that the effect of a transcription factor is determined principally by the location of its binding site in the promoter region. Thus, for example, CRP activates transcription when its binding site is centred near positions −61.5 or −41.5 upstream of the transcription start, but it interferes with RNAP binding, and therefore represses transcription, when the binding site is centred near positions −51.5 or −34.5.

Transcription activators

The majority of transcription activators bind to DNA sites located either upstream of, or overlapping, the target promoter −35 element [14]. Activators can function by either of two alternative modes of action: direct protein–protein interaction with RNAP, or introduction of a conformational change in the promoter DNA that makes it accessible to RNAP [15]. Both biochemical and genetic data suggest that most activators make direct contact with RNAP. In particular, the identification of mutant activator proteins that result in activation defects without affecting DNA binding (positive control mutants) has supported the idea that direct protein–protein interaction is the principal mechanism for transcription activation. The locations of the substitutions in these mutants define the region in the activator protein involved in contact with RNAP. In the same way, contact surfaces on RNAP have been identified from the location of substitutions that interfere with transcription activation without affecting factor-independent transcription initiation. Since the targets for activators can be located on different subunits of RNAP, it is possible to classify activators according to their target in RNAP.

Activators that contact αCTD

Many activators contact the αCTD [9]. The best characterized of these activators is CRP, but FNR, IHF, Fis, OxyR, OmpR, and CysB are also thought to interact with αCTD. The location for the binding site of these activators can vary considerably, from near position −91 to near position −41; however, it is always located upstream of the −35 sequence. This variability in the location of activators that contact αCTD is due to the presence of the flexible linker between the αNTD and αCTD [16]. Activation by these factors also requires DNA binding by the RNAP α subunits; the activator appears to facilitate docking of RNAP to the promoter via an αCTD–DNA interaction (Figure 2a). The consequence of this is that these activators primarily accelerate formation of the closed complex. Support for this comes from experiments showing that artificial protein–protein contacts can replace the CRP–αCTD interaction and activate transcription at the *lac* promoter [17]. Thus conformational changes in RNAP relayed by activator–αCTD contacts do not appear to play a major role.

Activators that contact the RNAP σ subunit

Together with αCTD, the σ^{70} subunit of RNAP is the major target of transcription activators. Significantly, these two RNAP segments both make specific contacts with the promoter DNA (−35 and −10 sequences for σ^{70} and the UP element for the α subunit) [16]. Activators known to interact with the σ^{70} subunit are PhoB, FNR and several bacteriophage proteins (λ cI protein, T4 MotA protein, and Mu Mor protein) [5]. Activators that contact the RNAP

σ^{70} subunit bind to DNA sites that overlap the target promoter -35 element (Figure 2b). Substitutions in σ^{70} that specifically affect activator-dependent transcription map between amino acids 570 and 580 and amino acids 591 and 613, located in region 4 close to the recognition motif for the -35 hexamer. Interestingly, a consensus-like -35 promoter element is usually absent. Thus protein–protein interaction between the activator and σ^{70} appears to substitute for the usual protein–DNA contact found at activator-independent promoters. This interaction can either recruit RNAP to the promoter region, similar to the role played by activators that contact αCTD, or enhance a subsequent step in transcription initiation. In the case of PhoB, initial recruitment of RNAP to the promoter region is likely to be the step affected by the activator, whereas the bacteriophage λ cI protein increases open-complex formation. It is noteworthy that several activators that contact σ^{70}, such as FNR and the Mu Mor protein, can interact simultaneously with αCTD (Figure 2c). It is likely that the two different interactions contribute to different steps in transcription initiation. The activator–αCTD interaction enhances recruitment of RNAP to the promoter, whereas the activator–σ interaction facilitates isomerization of the closed complex to the open complex [15].

Although most studies have focused on σ^{70}, the alternative σ subunits can also be targets for transcription activators. However, the extent of this kind of regulation is poorly understood at present. For example, activators such as Ada and CRP have been shown to contact RNAP containing σ^S, but the physiological significance of this is not known. It is likely that some activators are able to interact with similar contact sites in many different σ subunits.

Activators that contact other surfaces in RNAP

In principle, an activator could recruit RNAP to a promoter by contacting any surface, and a number of activators are known to contact RNAP surfaces other than αCTD and region 4 of σ^{70} [3]. Thus the DnaA protein activates the bacteriophage λ P_R promoter via interaction with the β subunit of RNAP. At CRP-dependent promoters, where the CRP binding site overlaps the promoter -35 sequence, CRP contacts αNTD, in addition to interacting with αCTD [12]. CRP contains two distinct surface-exposed regions that interact with αCTD and αNTD. Interestingly, at these promoters, the contact with αCTD is made by the upstream subunit of the CRP dimer and the contact with αNTD is made by the downstream subunit. The two contacts make distinct contributions to transcription activation. The CRP–αCTD interaction recruits RNAP to the promoter, whereas the CRP–αNTD interaction accelerates open-complex formation [12].

Activators that alter promoter DNA conformation

Several transcription activators appear to function primarily by altering the conformation of the target promoter DNA, without making direct contact with any of the RNAP subunits [15]. For example, some activators bind to targets upstream from the promoter elements and introduce a loop in the DNA that allows RNAP to contact distal DNA sequences, stabilizing its interaction with the promoter (Figure 3a). Another group of activators that function by altering DNA conformation are the members of the MerR family. For example, MerR, when triggered by mercuric ions, activates transcription of the *merP* gene (part of the mercury resistance locus of the Tn501 transposon). Unlike most activator-dependent promoters, the *merP* promoter carries -10 and -35 hexamer sequences that resemble the consensus. However, the spacing between these two elements (19 bp) is greater than the optimal spacing (17 bp) found in activator-independent promoters. Thus levels of transcription from the *merP* promoter in the absence of the activator are extremely low. However, when triggered by mercuric ions, MerR binds between the -10 and -35 hexamer elements at the *merP* promoter and induces a local twisting of the promoter DNA (Figure 3b). This results in repositioning of the -10 and -35 promoter elements so that efficient transcription initiation can take place.

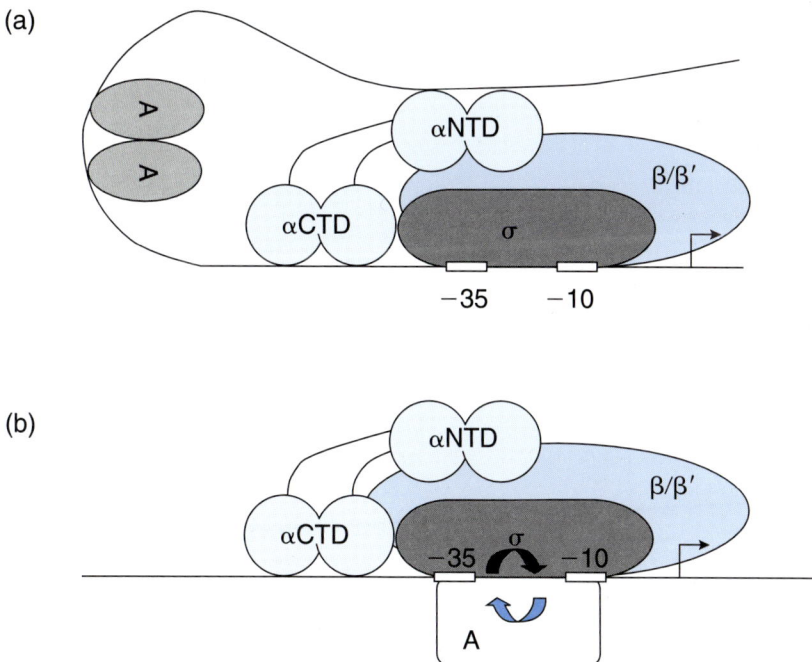

Figure 3. Activation of transcription by activator–promoter DNA interactions
(a) An upstream-bound activator (A) introduces a loop in the promoter DNA. This allows RNAP to contact distal promoter sequences, which stabilize its interaction with the promoter. (b) The -10 and -35 hexamers are repositioned due to local twisting of the promoter DNA induced by the binding of the activator (indicated by the arrows). RNAP can now bind and initiate transcription.

Transcription repressors

Many repressors prevent transcription by binding DNA at positions that interfere directly with RNAP binding [14]. Thus at many promoters subject to repression, operator sequences for a repressor are found to overlap or be immediately adjacent to the transcription start site (Figure 4a). A well studied example is the *E. coli lac* repressor, the first transcription regulator to be identified, which binds to a target site centred at position +11 with respect to the *lac* operon transcription start point. It has been found that the *lac* repressor, and many other repressors, function simply by interfering with the binding of RNAP to the target promoter. For factors that can act both as an activator or a repressor, it is the location of the binding site that determines their effect on transcription initiation.

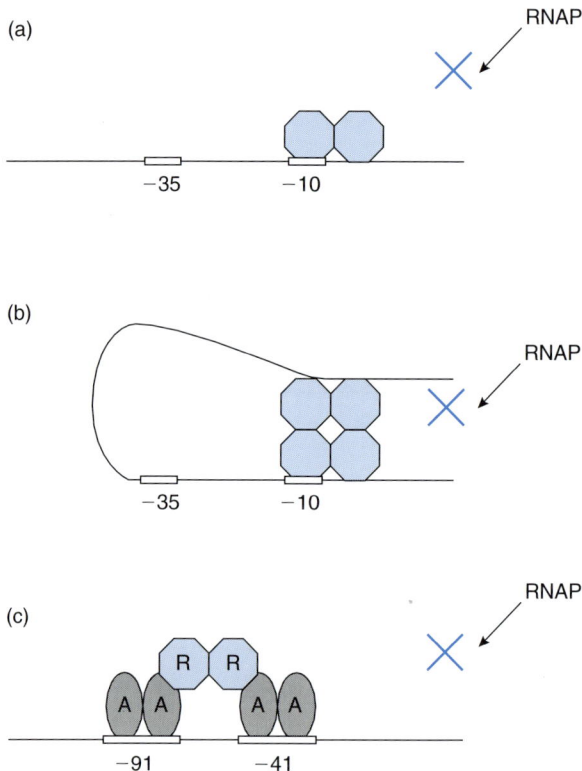

Figure 4. Mechanisms of transcriptional repression
In each diagram, the repressor is drawn as a dimer. (a) Repressor (octagon) blocks access of RNAP by binding to DNA and interfering directly with RNAP binding. (b) Distally bound repressors form a repression loop that hinders access of RNAP to the promoter. (c) Anti-activation mechanism: a repressor (R) binds to a tandemly bound activator (A) and prevents activation.

In many cases the mechanism of repression, although apparently simple, is unexpectedly complex. For example, for efficient repression of *lac* transcription, one of two other binding sites for the *lac* repressor is required in addition to the binding site at position +11 [18]. One of these supplementary sites is located at position −82 and the other is at +412. The presence of additional binding sites (operators) is a common theme in negative control of transcription initiation in bacteria – binding of more than one repressor molecule increases the local concentration of the repressor. Additionally, multiple binding sites can result in co-operative interactions that improve the affinity of the repressor for DNA, or in the formation of DNA micro-loops that hinder access of RNAP to the promoter (Figure 4b). Similar complexities are found during repression of the *gal* operon promoter by GalR, a homologue of the *lac* repressor. GalR binds two target sites centred at positions −60.5 and +53.5 and, together with the histone-like protein HU, creates a micro-loop around the *gal* promoter region [19]. Recent work has proposed that GalR does not prevent RNAP binding to the *gal* promoter, but that direct interaction between GalR and αCTD of RNAP maintains RNAP in an inactive state [20]. Thus, for full repression of *gal* transcription, GalR protein interferes both with the initial binding of RNAP to the promoter (via micro-loop formation) and with a subsequent step (via direct protein–protein interaction).

Transcription regulation by a repressor and an activator

Many naturally occurring bacterial promoters are extremely complex and most are regulated by more than one transcription factor. This permits the synthesis of a particular gene product to be controlled by more than one environmental stimulus; in effect, the promoter integrates information from different signals. For simplicity, we consider here only promoters that are regulated by two factors. Clearly there are three possible scenarios: regulation by two repressors, regulation by a repressor and an activator, and regulation by two activators. Although there are very few instances of promoters that are controlled by two repressors, there are many cases of regulation by a repressor and an activator. In most of these the repressor and the activator apparently function independently. For example, the *lac* promoter is active only when the *lac* repressor is removed from its target at position +11 (in response to lactose in the environment) and when the activator, CRP, binds to its target at position −61.5 (in response to low levels of glucose in the environment). However, there are cases where the functions of the repressor and the activator are not independent (Figure 4c). For example, promoters that control gene products needed for pyrimidine salvage are dependent on CRP for activity, but are repressed by CytR (a repressor whose activity is controlled by cytidine). At these promoters CytR binds directly to CRP and prevents CRP from activating transcription [21]. Thus CytR can be considered as an anti-activator rather than a simple repressor. Most of these promoters carry tandem

DNA sites for CRP separated by 52 base pairs. Interestingly, this arrangement is the optimal target for CytR, and hence CytR is targeted only to certain CRP-dependent promoters.

Transcription regulation by two activators

Many naturally occurring bacterial promoters are co-dependent on the action of two activators that make promoter activity contingent on two different physiological signals. Five different situations can account for co-dependence on two activators [22].

(i) The binding of one activator may be dependent on the binding of the other, and vice-versa (Figure 5a).

(ii) The binding of one activator may trigger the repositioning of the other, moving it from a location where it is unable to activate transcription to a location where it can activate transcription (Figure 5b).

(iii) The binding of one activator may alter the conformation of promoter DNA so that a second activator is able to make contact with RNAP (Figure 5c).

(iv) The two different activators may each make independent contacts with RNAP (Figure 5d).

(v) The second activator may counteract the action of a repressor that is interfering with the function of the first activator (Figure 5e).

Although examples of all five mechanisms have been found (see Figure 5), the independent contact mechanism (Figure 5d) appears to be the most common as it can be effected without interactions between the two different activators. In principle, this mechanism could accommodate as many activators as there are contact sites on the RNAP.

Perspectives

Over 5% of the genes of *E. coli* encode products concerned with transcriptional regulation and it is clear that a large variety of mechanisms are exploited. However, we can now propose a number of simple models to explain the majority of cases. These models hinge on our understanding of the structure of RNAP and of the role of the different RNAP subunits. Also, the realization that most bacterial transcription factors belong to a small number of families has greatly facilitated our understanding of different promoters. Immediate priorities for the future include the description, in structural terms, of the process of transcription initiation, elucidation of the precise mechanisms by which different activators and repressors function, and the structures at various activator–RNAP interfaces. Methods to address these priorities are now in place and progress should be rapid.

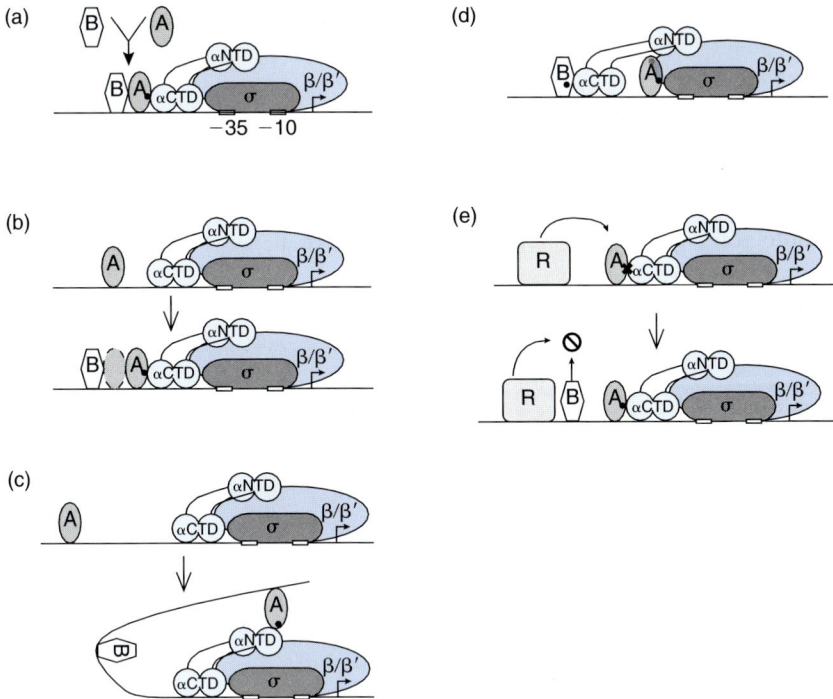

Figure 5. Mechanisms of co-dependence of a promoter on two activators

In each diagram, for simplicity, the activators are drawn as single subunits (in most cases they are likely to be dimers). (a) Co-operative binding of two activators. Binding of one activator (A) is dependent on the binding of a second activator (B) (e.g. MelR and CRP binding at the *E. coli melAB* promoter). (b) Repositioning. Binding of the second activator (B) triggers the repositioning of the first activator (A) from a position where it is unable to activate transcription to a position where it can activate transcription (e.g. CRP and MalT at the *E. coli malB* promoter region). (c) Alteration in DNA conformation. One activator (A) alters the conformation of the DNA to facilitate contacts between the other activator (A) and RNAP (e.g. at some σ^{54}-dependent promoters). (d) Independent contacts. The two activators make independent contacts with RNAP (e.g. CRP and FNR at the *E. coli ansB* promoter). (e) Counteraction of repression. One activator (B) counteracts a repressor (R) that would otherwise interfere with the activity of the other activator (A) (e.g. at the *E. coli nir* promoter).

Summary

- *Transcription initiation is the principal step at which bacterial gene expression is regulated. Bacterial transcription is due to a single multi-subunit RNA polymerase.*

- *The potential transcription initiation rate of any promoter is set by the efficiency with which RNA polymerase recognizes the different promoter sequence elements. The σ subunit plays the major role in the process of promoter recognition.*

- *Different RNA polymerase σ subunits can guide RNA polymerase to different promoters. The E. coli genome encodes seven different σ subunits, each of which allows the cell to respond to different environmental stimuli.*

- *A large number of transcription factors up-regulate and down-regulate expression from different promoters in response to environmental signals.*

- *Many transcription activators function by making a direct interaction with RNA polymerase. Some activators function by altering the conformation of promoter DNA.*

- *Most transcription repressors function by blocking access of RNA polymerase to their target promoter. In some cases, optimal repression depends on multiply bound repressor molecules that interact in complex ways.*

- *Many promoters are regulated by more than one transcription factor. A variety of mechanisms whereby a promoter can be regulated by a repressor and an activator, or by two activators, is known.*

References

1.   Muller-Hill, B. (1996) The lac Operon, Walter de Gruyter, Berlin
2.   Blattner, F.R., Plunkett, G., Bloch, C.A., Perna, N.T., Burland, V., Riley, M., Collado-Vides, J., Glasner, J.D., Rode, C.K., Mayhew, G.F., et al. (1997) The complete genome sequence of Escherichia coli K-12. Science 277, 1453–1474
3.   Ishihama, A. (1997) Promoter selectivity control of RNA polymerase In Nucleic Acids and Molecular Biology Vol. 11 (Eckstein, F. & Lilley, D., eds.), pp. 53–70, Springer-Verlag, Berlin
4.   Losick, R. & Chamberlin, M. J. (1976) RNA Polymerase, Cold Spring Harbor Laboratory Press, Cold Spring Harbor, NY
5.   Gross, C., Chan, C., Dombroski, A., Gruber, T., Sharp, M., Tupy, J. & Young, B. (1998) The functional and regulatory roles of sigma factors in transcription. Cold Spring Harbor Symp. Quant. Biol. 63, 141–155
6.   Ishihama, A. (1999) Modulation of the nucleoid, the transcription apparatus, and the translation machinery in bacteria for stationary phase survival. Genes Cells 4, 135–143
7.   Wosten, M. (1998) Eubacterial β factors. FEMS Microbiol. Rev. 22, 127–150
8.   Zhang, G., Campbell, E., Minakhin, L., Richter, C., Severinov, K. & Darst, S. (1999) Crystal structure of Thermus aquaticus RNA polymerase at 3.3 Å resolution. Cell 98, 811–824
9.   Ebright, R. & Busby, S. (1995) The Escherichia coli RNA polymerase α subunit: structure and function. Curr. Opin. Genet. Dev. 5, 197–203
10. Ross, W., Gosink, K., Salomon, J., Igarashi, K., Zou, C., Ishihama, A., Severinov, K. & Gourse, R. (1993) A third recognition element in bacterial promoters: DNA binding by the alpha subunit of RNA polymerase. Science 262, 1407–1413
11. Savery, N., Lloyd, G., Kainz, M., Gaal, T., Ross, W., Ebright, R., Gourse, R. & Busby, S. (1998) Transcription activation at Class II CRP-dependent promoters: identification of determinants in the C-terminal domain of the RNA polymerase α subunit. EMBO J. 17, 3439–3447
12. Niu, W., Kim, Y., Tau, G., Heyduk, T. & Ebright, R. (1996) Transcription activation at class II CAP-dependent promoters: two interactions between CAP and RNA polymerase. Cell 87, 1123–1134

13. Record, M., Reznikoff, W., Craig, M., McQuade, K. & Schlax, P. (1996) *Escherichia coli* RNA poly-
     merase, promoters and the steps of transcription initiation. In *Escherichia coli and Salmonella:*
     Cellular and Molecular Biology, 2nd edn (Niedhardt, F., et al., eds.), pp. 792–821, ASM Press,
     Washington DC

14. Gralla, J. & Collado-Vides, J. (1996) Organisation and function of transcription regulatory ele-
     ments, in *Escherichia coli and Salmonella: Cellular and Molecular Biology*, 2nd edn (Niedhardt, F., et
     al., eds.), pp. 1232–1245, ASM Press, Washington DC

15. Rhodius, V. & Busby, S. (1998) Positive activation of gene expression. *Curr. Opin. Microbiol.* **1**,
     152–159

16. Busby, S. & Ebright, R. (1994) Promoter structure, promoter recognition, and transcription acti-
     vation in procaryotes. *Cell* **79**, 743–746

17. Dove, S., Joung, J. & Hochschild, A. (1997) Activation of prokaryotic transcription through arbi-
     trary protein-protein contacts. *Nature (London)* **386**, 627–630

18. Muller-Hill, B. (1998) Some repressors of bacterial transcription. *Curr. Opin. Microbiol.* **1**, 145–151

19. Choy, H. & Adhya, S. (1996) Negative control, In *Escherichia coli and Salmonella: Cellular and*
     Molecular Biology, 2nd edn (Neidhardt, F., et al. eds.), pp. 1287–1299, ASM Press, Washington DC

20. Chatterjee, S., Zou, Y.-N., Roy, S. & Adhya, S. (1997) Interaction of Gal repressor with inducer
     and operator: induction of gal transcription from repressor-bound DNA. *Proc. Natl. Acad. Sci.*
     U.S.A. **94**, 2957–2962

21. Valentin-Hansen, P., Sogaard-Andersen, L. & Pedersen, H. (1996) A flexible partnership: the CytR
     anti-activator and the cAMP-CRP protein, comrades in transcription control. *Mol. Microbiol.* **20**,
     461–466

22. Busby, S. & Savery, N. (1999) Transcription activation at bacterial promoters. In *The Embryonic*
     Encyclopaedia of Life Sciences. www.els.net/elsonline/html/A0000855.html, Nature Life Sciences,
     London

3

Regulation of the initiation of eukaryotic transcription

Grace Gill[1]

Department of Pathology, Harvard Medical School, 200 Longwood Avenue, Boston, MA 02115, U.S.A.

Introduction

Transcription is the fundamental process whereby the genetic information encoded in the DNA is first expressed. During development, expression of specific genes is turned on or off at precise times and in particular cells, giving rise to the diversity and specificity of cell function. Transcriptional regulation also enables cells to respond to environmental cues such as the availability of nutrients or viral infection. Thus failure to properly regulate transcription can lead to severe developmental abnormalities or disease. Regulation of gene expression is determined in large part by the activity of transcriptional activator proteins that bind specific DNA sequences near the gene. The first step in transcribing a protein-coding gene, i.e. the binding of the RNA polymerase II machinery to the promoter, is subject to extensive regulation. In general, wrapping of the DNA around nucleosomes and packaging into higher-order chromatin inhibits binding by the transcription machinery. Therefore chromatin-modifying complexes, which can either enhance or relieve chromatin-mediated repression, are critical to regulated transcription (see Chapter 4 in this volume by A.P. Wolffe). This chapter will focus on the factors that influence the efficiency of binding by the transcription machinery to DNA that has already been made accessible, namely the DNA sequence of the core promoter and interactions with transcriptional activators and co-activators. Synergistic activation by several transcription factors working

[1]*e-mail: grace_gill@hms.harvard.edu*

together provides additional specificity and potency to gene regulation. As described below, studies on the functional interactions of the RNA polymerase II transcription machinery with promoters and activators have provided new insights into the mechanisms that regulate the initiation of eukaryotic transcription.

A typical eukaryotic promoter

Several functionally distinct types of DNA elements that can contribute to gene regulation in eukaryotes have been identified. These include the core promoter, regulatory promoter and enhancer sequences. Additional sequences may also contribute to regulation of gene expression in chromatin. The core promoter serves as a binding site for the RNA polymerase II transcription machinery and determines both the start site and the direction of transcription. The core promoter is sufficient to support correctly initiated transcription *in vitro*. Common core promoter elements include the TATA box, which is generally located 25–30 bp upstream of the start site, the initiator (Inr), which flanks the start site, and the downstream promoter element (DPE) found approximately 30 bp downstream of the start site (Figure 1) [1,2]. Core promoters can include a TATA box alone, an Inr element alone or both together. The DPE is found in combination with an Inr at some promoters lacking TATA boxes. The sequence of the core promoter not only determines the basal, or unregulated, level of transcription, but may also contribute to regulation of the gene [3,4].

Although the core promoter is sufficient to support correctly initiated transcription of a gene *in vitro*, expression of most genes *in vivo* is dependent on additional regulatory sequences. Important regulatory sequences located within

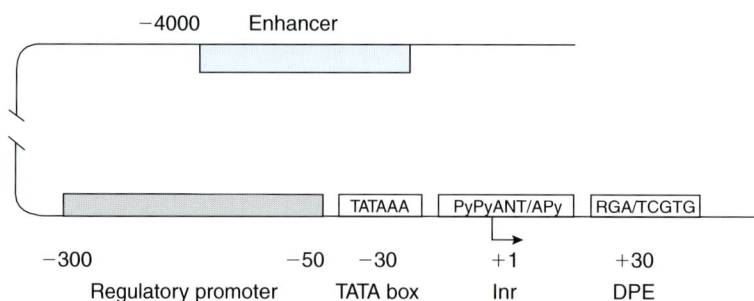

Figure 1. Typical promoter elements that regulate transcription of a gene in higher eukaryotes
The consensus sequences of the common core promoter elements and their positions relative to the start site of transcription are shown. Both the regulatory promoter and the enhancer are composed of binding sites for multiple transcription factors. The actual sequence, size and position of these elements varies greatly among different promoters. Py denotes a pyrimidine nucleotide.

a few hundred base pairs 5' of the core promoter are often referred to as the regulatory promoter. Enhancers, which are functionally defined as DNA elements that increase transcription of a gene in a position- and orientation-independent manner, can be located at great distances either upstream or downstream of the core promoter. Binding sites for many transcription factors are found in both regulatory promoters and enhancers, and the distinction between these transcriptional control elements is sometimes blurred. Detailed analysis of enhancers such as the interferon-β enhancer has revealed that these DNA sequences serve as a template for the assembly of a precisely configured DNA–protein complex termed the enhanceosome [5]. As described below, co-operative protein–protein interactions between many activators bound to the enhancer contribute to both the specificity and potency of enhancer-mediated transcriptional activation.

The RNA polymerase II machinery

RNA polymerase II requires a number of additional factors to specifically initiate transcription of a gene, dependent on the promoter sequences. One of the great advances in transcription research in recent years has been the identification and cloning of these general transcription factors (GTFs). The GTFs include TFIIA, TFIIB, TFIID, TFIIE, TFIIF and TFIIH [6]. TFIID is the only GTF that binds specifically and independently to the core promoter. Both the TATA-box-binding protein (TBP) and TBP-associated factors (TAF$_{II}$s) in the multiprotein TFIID complex contribute to binding the core promoter. TBP binds the TATA box, TAF$_{II}$250 and TAF$_{II}$150 together bind to the Inr element, and TAF$_{II}$60 contributes to binding to the DPE [2,6,7]. Interaction of TFIIB with DNA just 5' of the TATA box can also influence binding of the transcription machinery to the core promoter [8]. Using highly purified components, researchers have defined an order of assembly of the GTFs on a promoter. TFIID, together with TFIIA, binds first to the promoter, followed by TFIIB, an RNA polymerase II–TFIIF complex, then TFIIE and lastly TFIIH [6,9]. Recent studies have suggested, however, that many of these factors bind each other off the DNA, leading to a simplified two-step reaction in which TFIID binds together with TFIIA, followed by binding of a large pre-assembled complex including RNA polymerase II and the remaining GTFs (Figure 2) [6]. This form of the polymerase, termed the holoenzyme, is also associated with accessory factors, such as SRB (suppressor of RNA polymerase B) and mediator proteins in yeast and p300/CBP [cAMP-response-element-binding protein (CREB)-binding protein] in humans. Although these factors are not required for basal transcription *in vitro*, they have been found to contribute to regulated gene expression [10]. These studies have revealed a stunning complexity in the eukaryotic transcription machinery, with over 50 proteins assembled at the core promoter.

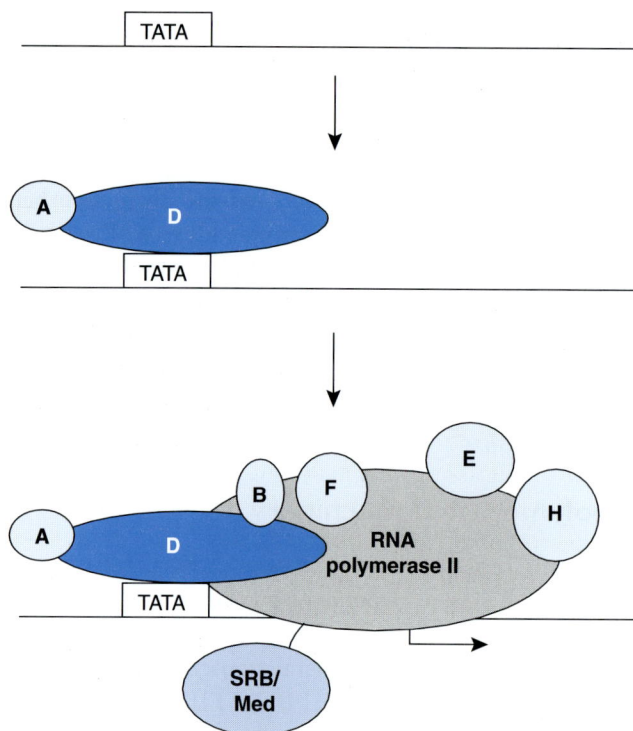

Figure 2. General transcription factors (GTFs) may assemble on the core promoter in as few as two steps: binding of TFIIA/TFIID and then binding of a holoenzyme complex containing GTFs, RNA polymerase II, and regulatory proteins
For simplicity, the core promoter shown has a TATA box. The SRB/mediator (Med) proteins, which associate with the holoenzyme in yeast, are used here to represent the many regulatory proteins that associate with the holoenzyme in yeast and humans. A, TFIIA; B, TFIIB etc.

Biochemical studies using model templates such as the adenovirus major late promoter led to the identification of the GTFs required to support specific initiation of transcription by RNA polymerase II [6]. As more promoters are analysed, however, it has become clear that the same set of GTFs may not function at all promoters. The TAF$_{II}$s in the TFIID complex, for example, are required for transcription from a core promoter lacking a TATA box *in vitro*, although TBP alone is sufficient to support transcription from a promoter with a consensus TATA box [11]. Studies in yeast confirm that the requirement of TAF$_{II}$s for transcription *in vivo* is dependent upon the sequence of the core promoter [12]. Similarly, TFIIE has been found to be dispensable for transcription of certain promoters *in vitro* and *in vivo* [13,14]. In fact, recent studies have revealed a surprising diversity of GTFs, including two TBP-related factors, at least one tissue-specific TAF$_{II}$ and a testis-specific TFIIA isoform, further supporting the notion that distinct sets of factors may assemble at specific promoters [15–18]. Although the mechanisms that determine which basal transcription factors are required at a given promoter are not well understood, the exact composition of the transcription machinery that assembles at the promoter is likely to contribute to regulation.

Activators and co-activators

Regulated expression of a gene depends in large part on the activity of transcription factors that bind to specific sequences in the regulatory promoter or enhancer. Many of the dynamic changes in gene expression in response to a developmental programme or external signal are due to changes in the levels or activity of specific transcriptional activators and repressors. Although the activity of promoter-specific repressors is important for eukaryotic gene regulation, repression mechanisms will not be discussed here. Transcriptional activators have at least two functional domains: a DNA-binding domain and an activation domain [19]. The DNA-binding domain binds a specific sequence present in the regulatory promoter or enhancer, and thereby serves to localize the activator near the gene to be regulated. The activation domain mediates protein–protein interactions with other activators, co-activators and GTFs to increase transcription. Although the first activation domains to be identified were noticeably rich in acidic amino acids, many different amino acid sequences have been found to function as activation domains [20]. Studies of several transcriptional activators, in particular the VP16 protein from herpes simplex virus, which does not bind DNA directly but is localized to the promoter by binding to a DNA-binding protein, have identified interactions between activation domains and the GTFs TFIIA, TBP, TFIIB, TFIIF and TFIIH [20]. Although it has not been established that all of the observed interactions are physiologically relevant, it is likely that activator interactions with many different components of the RNA polymerase II transcriptional machinery can enhance transcription. Thus, there are many targets for activation domains, and, in fact, a single activator can interact with multiple general transcription factors or co-activators to stimulate transcription [21].

Surprisingly, when researchers analysed transcription *in vitro* using purified activators, GTFs and RNA polymerase II, it was found that these components were not sufficient to generate high levels of activated transcription [22]. These *in vitro* studies revealed that additional co-activators were required in addition to the basal transcription machinery and sequence-specific activators. Subsequent biochemical and genetic studies have identified numerous co-activators that contribute to activated transcription [21]. Many co-activators have been shown to bind directly to activators, providing a mechanism for targeting co-activator function to specific promoters. For example, the interaction of $TAF_{II}110$ with the glutamine-rich activation domains of the transcriptional activator Sp1 is important for Sp1-mediated activation [23]. Similarly, the thyroid-hormone-receptor-associated protein 220 component of the human mediator complex binds directly to the vitamin D and thyroid hormone receptors in a ligand-dependent manner [24]. Some co-activators, such as the TAF_{II}s in

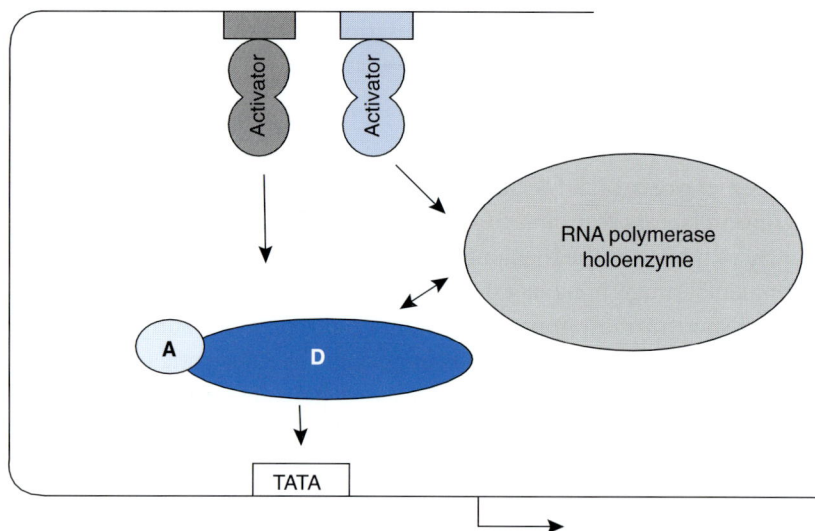

Figure 3. Transcriptional activators interact with components of the TFIIA/TFIID or holoenzyme complexes to increase recruitment of the RNA polymerase transcription machinery to the promoter
The DNA is shown looping to accommodate the many protein–protein and protein–DNA interactions indicated by arrows. The interaction of two activators with different components of the transcription machinery leads to synergistic activation. Co-activators (not shown) contribute to activation by bridging between activators and GTFs or by other mechanisms. Abbreviations are as for Figure 2.

the TFIID complex and holoenzyme components such as the SRB proteins in yeast or p300/CBP in mammals, are tightly associated with the general transcription machinery. These co-activators may function, in part, as simple bridges to link activators with the RNA polymerase II transcriptional machinery. It should be noted that some co-activators, such as CBP, which contribute to activation on simple DNA templates also have nucleosome-modifying activities that are likely to contribute to activation on chromatin templates [10]. Recent studies have raised the exciting possibility that, similar to activator proteins, the level or activity of co-activators may be modified in response to extracellular signals, leading to a co-ordinated change in expression of the many genes that are dependent on a particular co-activator [14].

Mechanisms of activation

Multiple steps are required to produce a correctly initiated full-length transcript. These steps include binding of the GTFs and RNA polymerase II to the promoter, melting of the template strands, formation of the first phosphodiester bond (initiation), release of the RNA polymerase II complex from the promoter (promoter clearance), movement of the polymerase through the gene (elongation) and termination. In addition, the components of the transcription machinery are recycled to allow multiple rounds of

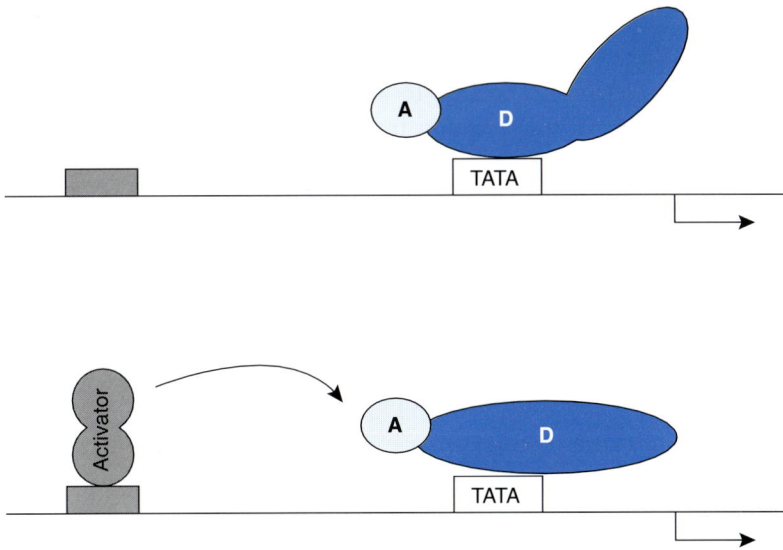

Figure 4. Activator interactions may stabilize a conformation of the TFIIA–TFIID–DNA complex that recruits the holoenzyme more efficiently
Abbreviations are as for Figure 2.

transcription from a single template. Each of these steps could, in principle, be subject to negative regulation by repressors or positive regulation by activators.

It is perhaps not surprising that the first step in this process, binding of the RNA polymerase II transcription complex to the promoter, is subject to extensive regulation. As shown in Figure 2, binding of the RNA polymerase II machinery to the promoter is thought to occur in at least two steps — binding of TFIID together with TFIIA to the core promoter, and subsequent binding of an RNA polymerase II holoenzyme complex to the resulting DNA–protein assembly. Although there are clear examples of activators that stimulate elongation, many transcriptional activators act to stimulate recruitment of the RNA polymerase II machinery to the promoter (Figure 3) [19]. Many activators, when bound to nearby sites, function to increase TFIID binding to the core promoter *in vitro* and *in vivo* [9,25]. Studies using immobilized template assays to analyse proteins bound at the promoter have also shown that, under conditions where TBP/TFIID is bound at the promoter, activators can increase recruitment of holoenzyme components, including TFIIB and RNA polymerase II [26,27]. Activators function to increase binding of the transcription machinery to the promoter in at least two ways: (i) the additional affinity provided by a simple protein contact and (ii) stabilization of a conformation that increases assembly of the GTFs on the promoter (Figure 4).

The additional affinity provided by binding to an activator localized at the promoter is, in many cases, sufficient to increase binding of the transcription machinery to the core promoter [19]. Experiments using fusion proteins dramatically illustrate that increasing the local concentration of the transcription machinery through many diverse types of interactions results in increased transcription. Direct fusion of TBP to a heterologous DNA-binding domain produces high levels of transcription from promoters with the appropriately positioned binding sites [28,29]. Similarly, a heterodimerization domain, when localized to a promoter by fusion to a DNA-binding domain, was shown to activate transcription when the corresponding dimerization domain was fused to TFIIB [30]. Artificial recruitment of several other holoenzyme components also leads to activated levels of transcription [19]. The results of such 'activator bypass' experiments argue that the activator does not have to interact in any special way with the transcription machinery, since an activation domain can easily be substituted by a DNA-binding domain or a dimerization domain. Consistent with the idea that many activators increase the affinity of the transcription machinery for the promoter, but do not otherwise modify its activity, increasing the concentration of holoenzyme in an *in vitro* reaction is sufficient to obtain activated levels of transcription [31].

Interaction with an activator has also been found to promote a conformational change that enhances binding of the transcription machinery to the promoter. Studies of the mechanisms of transcriptional activation by the Zebra activator from Epstein–Barr virus have revealed that Zebra interacts with TFIID. In this case, however, increasing the concentration of TFIIA–TFIID to ensure binding to the promoter was not sufficient to obtain activated levels of transcription. Interaction with the Zebra activator stabilizes a conformation of the TFIIA–TFIID–TATA complex with increased affinity for the RNA polymerase II holoenzyme [32]. Although the exact nature of this conformational change is not known, other activators have been observed to alter the interaction of TFIID with the core promoter, as judged by DNaseI footprinting, suggesting that the conformation of TFIID may be commonly subject to regulation [9]. Interestingly, the pattern of the TFIID footprint varies on different core promoters, raising the possibility that the sequence of the core promoter may influence the conformation of TFIID and thereby the response to certain activators.

Transcriptional synergy

In eukaryotes, different combinations of a relatively small number of transcription factors are used to generate a large array of specific transcription patterns in a process often referred to as combinatorial control. The effect of two or more activators working together at a promoter is generally much more than additive [5]. Synergistic activation has important consequences for the specificity of gene regulation, since it ensures that a gene will only be expressed when each of the activators that regulate it are functional. For example, induction of the gene encoding interferon-β in response to virus infection requires the transcription factor nuclear factor κB (NF-κB). Not all signals that activate NF-κB, however, result in stimulation of interferon-β expression. Interferon-β transcription is induced only when all of the transcription factors, including NF-κB, activating transcription factor 2 (ATF2)/c-Jun and interferon regulatory factor-3 (IRF-3), function together at the enhancer [33,34]. Thus the interferon-β enhancer serves to integrate different external signals. Several mechanisms are likely to contribute to synergistic activation [5]. First, interactions between activators may allow for co-operative binding and increased stability of the activator–DNA complex. Secondly, non-competitive interaction of multiple activators with different surfaces of the transcriptional machinery should increase the effective interaction energy. Thirdly, if multiple activators stimulate different steps in the transcription reaction, such as recruitment of TFIID and the holoenzyme, the effects may be greater than additive. It is likely that all of these mechanisms of synergy contribute to the high levels of activation mediated by enhancers.

Perspectives

Although studies of several model promoters and activators has revealed much about regulated transcription initiation in eukaryotes, several questions remain to be answered. In particular, regulation of very few cellular genes has been studied in sufficient detail to understand the complex interplay between the sequence of the core promoter, the requirement for individual GTFs and the activity of specific activators. In addition, many co-activators have been identified, including the multi-subunit human mediator complex, and the mechanisms by which these co-activators function with activators to regulate transcription at specific promoters continues to be a subject of active investigation. It is interesting to note that several co-activators have associated kinase or acetyltransferase activities, which may contribute to their function. Since gene regulation *in vivo* occurs in the context of chromatin, it is of significant interest to understand how the functions of activators, co-activators and GTFs described here are co-ordinated with the activity of chromatin-modifying complexes. Further studies of transcriptional activation mechanisms should advance our knowledge of how transcription of each of the more than 60000 human genes is properly regulated.

Summary

- *DNA sequences that determine transcriptional regulation of a typical eukaryotic gene consist of a core promoter, which serves as a binding site for the GTF TFIID, and regulatory promoter or enhancer sequences, which bind transcriptional activators.*

- *The RNA polymerase II transcription machinery consists of over 50 proteins which are thought to bind to the core promoter in as few as two steps: binding of TFIIA–TFIID, followed by binding of a large pre-assembled holoenzyme complex consisting of the remaining GTFs, RNA polymerase II and associated regulatory proteins.*
- *Activators function to increase binding of the transcription machinery to the promoter in at least two ways: (i) simple protein–protein interactions with activators increases the affinity of the transcription machinery for the promoter, and (ii) some activators stabilize a conformation of the TFIIA–TFIID–DNA complex that enhances binding of the holoenzyme.*
- *Recent studies have identified many co-activators that function with activators to increase transcription by the RNA polymerase II transcription machinery. Although some co-activators may serve as bridges to connect activators with the transcription machinery, the mechanism of action of many co-activators has not yet been determined.*

References

1. Smale, S.T. (1997) Transcription initiation from TATA-less promoters within eukaryotic protein-coding genes. *Biochim. Biophys. Acta* **1351**, 73–88
2. Burke, T.W. & Kadonaga, J.T. (1997) The downstream core promoter element, DPE, is conserved from Drosophila to humans and is recognized by TAFII60 of Drosophila. *Genes Dev.* **11**, 3020–3031
3. Harbury, P.A. & Struhl, K. (1989) Functional distinctions between yeast TATA elements. *Mol. Cell. Biol.* **9**, 5298–5304
4. Ohtsuki, S., Levine, M. & Cai, H.N. (1998) Different core promoters possess distinct regulatory activities in the Drosophila embryo. *Genes Dev.* **12**, 547–556
5. Carey, M. (1998) The enhanceosome and transcriptional synergy. *Cell* **92**, 5–8
6. Orphanides, G., Lagrange, T. & Reinberg, D. (1996) The general transcription factors of RNA polymerase II. *Genes Dev.* **10**, 2657–2683
7. Chalkley, G.E. & Verrijzer, C.P. (1999) DNA binding site selection by RNA polymerase II TAFs: a TAF(II)250-TAF(II)150 complex recognizes the initiator. *EMBO J.* **18**, 4835–4845
8. Lagrange, T., Kapanidis, A.N., Tang, H., Reinberg, D. & Ebright, R.H. (1998) New core promoter element in RNA polymerase II-dependent transcription: sequence-specific DNA binding by transcription factor IIB. *Genes Dev.* **12**, 34–44
9. Roeder, R.G. (1996) The role of general initiation factors in transcription by RNA polymerase II. *Trends Biochem. Sci.* **21**, 327–335
10. Parvin, J.D. & Young, R.A. (1998) Regulatory targets in the RNA polymerase II holoenzyme. *Curr. Opin. Genet. Dev.* **8**, 565–570
11. Pugh, B.F. & Tjian, R. (1991) Transcription from a TATA-less promoter requires a multisubunit TFIID complex. *Genes Dev.* **5**, 1935–1945
12. Shen, W.C. & Green, M.R. (1997) Yeast TAF(II)145 functions as a core promoter selectivity factor, not a general coactivator. *Cell* **90**, 615–624
13. Parvin, J.D., Timmers, H.T. and Sharp, P.A. (1992) Promoter specificity of basal transcription factors. *Cell* **68**, 1135–1144

14. Holstege, F.C., Jennings, E.G., Wyrick, J.J., Lee, T.I., Hengartner, C.J., Green, M.R., Golub, T.R., Lander, E.S. & Young, R.A. (1998) Dissecting the regulatory circuitry of a eukaryotic genome. *Cell* **95**, 717–728

15. Dikstein, R., Zhou, S. & Tjian, R. (1996) Human TAFII 105 is a cell type-specific TFIID subunit related to hTAFII130. *Cell* **87**, 137–146

16. Hansen, S.K., Takada, S., Jacobson, R.H., Lis, J.T. & Tjian, R. (1997) Transcription properties of a cell type-specific TATA-binding protein, TRF. *Cell* **91**, 71–83

17. Dantonel, J.C., Wurtz, J.M., Poch, O., Moras, D. & Tora, L. (1999) The TBP-like factor: an alternative transcription factor in metazoa? *Trends Biochem. Sci.* **24**, 335–339

18. Ozer, J., Moore, P.A. & Lieberman, P.M. (2000) A testis-specific transcription factor IIA (TFIIAtau) stimulates TATA-binding protein-DNA binding and transcription activation. *J. Biol. Chem.* **275**, 122–128

19. Ptashne, M. & Gann, A. (1997) Transcriptional activation by recruitment. *Nature (London)* **386**, 569–577

20. Triezenberg, S. (1995) Structure and function of transcriptional activation domains. *Curr. Opin. Genet. Dev.* **5**, 190–196

21. Berk, A.J. (1999) Activation of RNA polymerase II transcription. *Curr. Opin. Cell Biol.* **11**, 330–335

22. Gill, G. & Tjian, R. (1992) Eukaryotic coactivators associated with the TATA box binding protein. *Curr. Opin. Genet. Dev.* **2**, 236–242

23. Gill, G., Pascal, E., Tseng, Z. & Tjian, R. (1994) A glutamine-rich hydrophobic patch in transcription factor Sp1 contacts the dTAF$_{II}$110 component of the Drosophila TFIID complex and mediates transcriptional activation. *Proc. Natl. Acad. Sci. U.S.A.* **91**, 192–196

24. Yuan, C.X., Ito, M., Fondell, J.D., Fu, Z.Y. & Roeder, R.G. (1998) The TRAP220 component of a thyroid hormone receptor-associated protein (TRAP) coactivator complex interacts directly with nuclear receptors in a ligand-dependent fashion. *Proc. Natl. Acad. Sci. U.S.A.* **95**, 7939–7944

25. Kuras, L. & Struhl, K. (1999) Binding of TBP to promoters in vivo is stimulated by activators and requires Pol II holoenzyme. *Nature (London)* **399**, 609–613

26. Choy, B. and Green, M.R. (1993) Eukaryotic activators function during multiple steps of preinitiation complex assembly. *Nature (London)* **366**, 531–536

27. Ranish, J.A., Yudkovsky, N. & Hahn, S. (1999) Intermediates in formation and activity of the RNA polymerase II preinitiation complex: holoenzyme recruitment and a postrecruitment role for the TATA box and TFIIB. *Genes Dev.* **13**, 49–63

28. Chatterjee, S. & Struhl, K. (1995) Connecting a promoter-bound protein to TBP bypasses the need for a transcriptional activation domain. *Nature (London)* **374**, 820–822

29. Klages, N. & Strubin, M. (1995) Stimulation of RNA polymerase II transcription initiation by recruitment of TBP in vivo. *Nature (London)* **374**, 822–823

30. Gonzalez-Couto, E., Klages, N. & Strubin, M. (1997) Synergistic and promoter-selective activation of transcription by recruitment of transcription factors TFIID and TFIIB. *Proc. Natl. Acad. Sci. U.S.A.* **94**, 8036–8041

31. Gaudreau, L., Adam, M. & Ptashne, M. (1998) Activation of transcription in vitro by recruitment of the yeast RNA polymerase II holoenzyme. *Mol. Cell* **1**, 913–916

32. Chi, T. & Carey, M. (1996) Assembly of the isomerized TFIIA–TFIID–TATA ternary complex is necessary and sufficient for gene activation. *Genes Dev.* **10**, 2540–2550

33. Thanos, D. & Maniatis, T. (1995) Virus induction of human IFN beta gene expression requires the assembly of an enhanceosome. *Cell* **83**, 1091–1100

34. Wathelet, M.G., Lin, C.H., Parekh, B.S., Ronco, L.V., Howley, P.M. & Maniatis, T. (1998) Virus infection induces the assembly of coordinately activated transcription factors on the IFN-beta enhancer in vivo. *Mol. Cell* **1**, 507–518

4

Transcriptional regulation in the context of chromatin structure

Alan P. Wolffe[1]

Laboratory of Molecular Embryology, National Institute of Child Health and Human Development, NIH, Bethesda, MD 20892-5431, U.S.A.

Introduction

Biophysicists and biochemists have often focused on chromatin structure simply as an engineering problem. The highly conserved histones are thought to wrap DNA around themselves to assemble monotonous arrays of nucleosomes, which in turn self-associate to form semi-crystalline higher-order structures. This is very convenient for scientists who require structurally uniform samples to interpret their experiments on the physical properties of biological material. However, this homogeneous packaging of DNA within the chromosome is not consistent with the experience of those molecular geneticists and cell biologists who investigate chromatin function. These investigators find the organization of DNA within the chromosome to be highly varied, reflecting the functional differentiation of the chromosome into distinct domains.

Recent observations provide a molecular explanation that reconciles the dual requirements of packaging of DNA into chromatin via nucleosomal structures and the functional differentiation of the chromosome into distinct

[1]*Address for correspondence: Sangamo BioSciences Inc., Point Richmond Tech Center, 501 Canal Blvd., Suite A100, Richmond, CA 94804, U.S.A. (e-mail: awolffe@sangamo.com).*

domains. The assembly of variants of the histones themselves into chromatin, and the inclusion of specific *trans*-acting factors with striking similarities to the histones, can retain some of the architectural features of canonical nucleosomes, yet provide highly differentiated structures necessary to facilitate transcription. Moreover, chromatin structure is dynamic, accommodating the various functions that require DNA as a template. In the past, changes in chromatin conformation were usually attributed to DNA being used for some purpose, but it is now clear that alterations in chromatin conformation can have a causal role in gene control. Recent progress in the identification and characterization of chromatin-modifying proteins, particularly acetyltransferases, deacetylases and members of the SWI/SNF (mating type switching/sucrose non-fermenting in *Saccharomyces cerevisiae*) superfamily of proteins, provides mechanistic insight into exactly how chromatin remodelling is targeted to regulate transcription. Evolution has been remarkably successful in shaping chromatin such that it can become alternately transparent or opaque, facilitating or restricting the access of transcription factors and RNA polymerase to DNA. The molecular mechanisms controlling this access are central to gene regulation.

Principles of gene regulation in chromatin

Metazoan genes are regulated by a series of *cis*-acting elements and *trans*-acting factors. The *cis*-acting elements include locus control regions, enhancers and proximal promoter elements (Figure 1a). These are each assembled into precise nucleoprotein architectures that communicate with each other and with the local chromatin environment (see later) and which can promote transcription at several levels (Figure 1b). The formation of chromatin loops including 10–100 kb of DNA can facilitate interactions between locus control regions and enhancers or promoters. Higher-order chromatin structure can promote interactions over the 1–10 kb range, and positioned nucleosomes can allow enhancers or promote proximal regulatory elements to interact with the basal transcriptional machinery. For example, an oestrogen-responsive enhancer in the *Xenopus* vitellogenin promoter lies 300 bp upstream of the TATA box and 180 bp upstream of liver-specific proximal promoter element. Nucleosome assembly on the DNA between the enhancer and the proximal promoter element stimulates transcription more than 100-fold when compared with a non-specific chromatin structure (Figure 2). Histones can also repress transcription by preventing general transcription factors and the TATA box from recruiting RNA polymerase (see later). Sequence-specific transcription factors bound at locus control regions, enhancers and promoters mediate the recruitment of a complex chromatin remodelling machinery that controls the interplay of histones, transcription factors and RNA polymerase at the TATA box through both positive and negative effects.

(a)

(b)

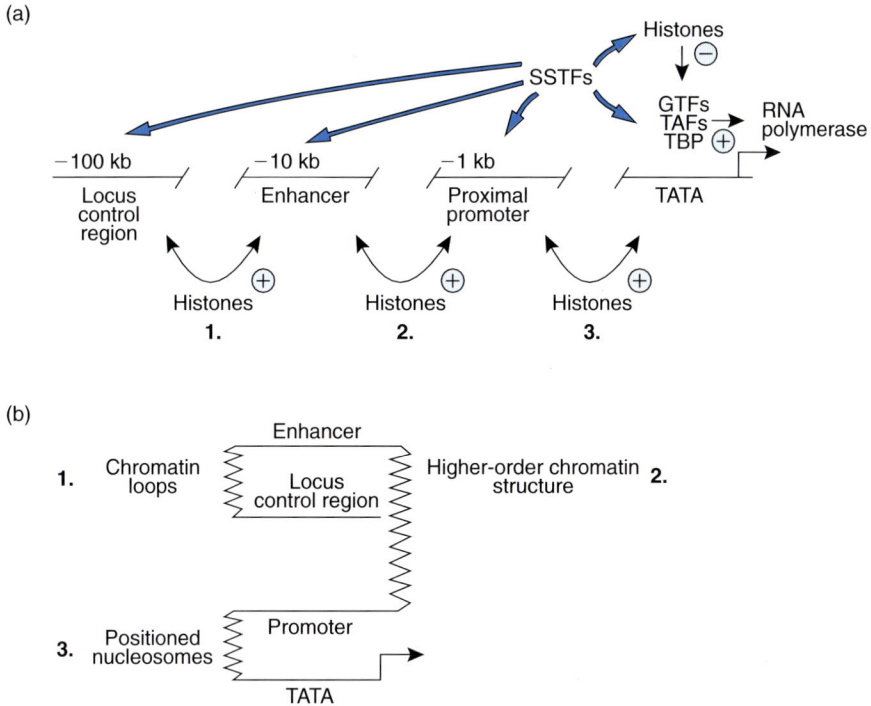

Figure 1. Regulation of eukaryotic gene expression
(a) Three types of *cis*-acting elements exist, arbitrarily defined by their distance from the site of TATA-box-binding protein (TBP) binding at the TATA box. These are locus control regions, enhancers and promoter proximal elements, at which sequence-specific transcription factors (SSTFs) bind. Communication between transcription factors bound at these elements may be facilitated by packaging of the intervening DNA by histones. SSTFs influence the binding of general transcription factors (GTFs), TBP and TBP-associated factors (TAFs). The TAFs, GTFs, and TBP recruit RNA polymerase to the promoter and facilitate transcription. They may also direct the displacement of histones from the TATA box. (b) Several levels of nucleoprotein higher-order structure also allow communication between the *trans*-acting factors bound to *cis*-acting sequences.

Core histones and related regulatory proteins

The four core histones (H2A, H2B, H3 and H4) have been highly conserved throughout eukaryotic evolution. Each has a very similar C-terminal domain structure, known as the histone-fold, that directs the formation of specific heterodimers between the histones and also determines the path of DNA in the nucleosome [2,3]. Each histone-fold contains a long central helix bordered on each side by a loop segment and a shorter helix. The long central helix acts as a heterodimerization interface. Dimerization creates three DNA-binding surfaces through the interaction of loop segments at each end of the long central helix and through the juxtaposition of the two short α-helices flanking the N-termini of the long central helix.

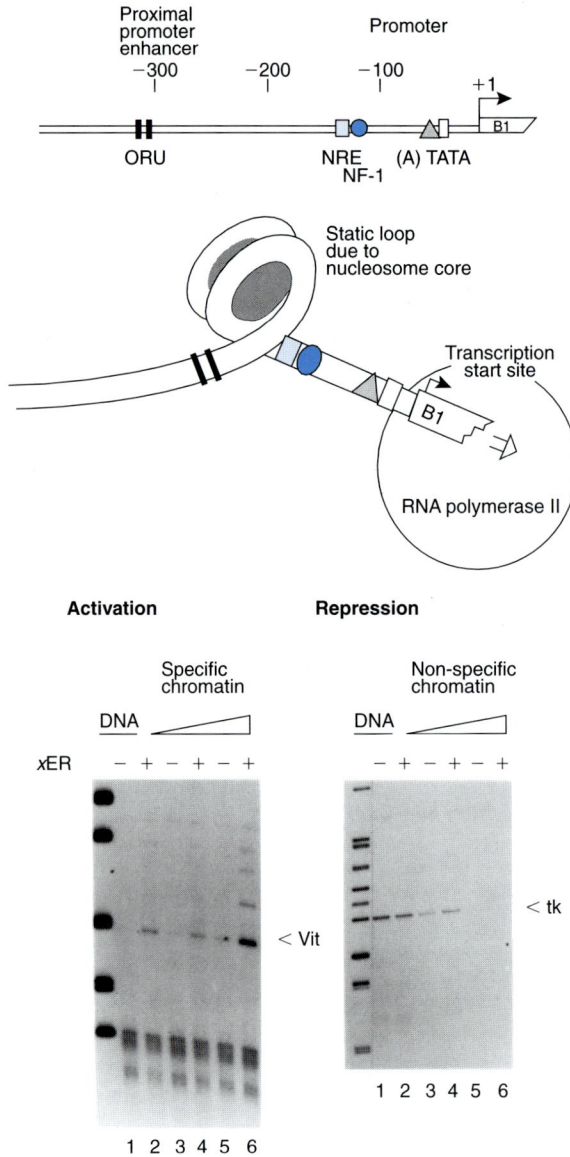

Figure 2. Role of a static loop created by a nucleosome including part of the *Xenopus* vitellogenin B1 promoter in potentiating transcription

From a specific chromatin template that allows positioned assembly of nucleosomes, transcription increases with increasing numbers of nucleosomes whereas transcription from a non-specific chromatin template in which nucleosomes are not positioned decreases with increasing numbers of nucleosomes. ORU, oestrogen response unit; NRE, negative regulatory element; NF-1, nuclear factor-1; Vit, vitellogenin mRNA; tk, thymidine kinase mRNA; xER, *Xenopus* oestrogen receptor. Reproduced from Schild, C., Claret, F.-X., Wahli, W. & Wolffe, A.P. A nucleosome-dependent static loop potentiates estrogen-regulated transcription from the *Xenopus* vitellogenin B1 promoter *in vitro*. *EMBO J.* **12**, 423–433 © (1993), with permission from the European Molecular Biology Organization.

The N-terminal tails of the core histones and the C-terminal tail of histone H2A protrude on the outside of the nucleosome, where they can potentially contact other nucleoprotein complexes, leading to activation or repression of transcription in a promoter-specific manner. The N-terminal tail domains are also the targets of specific enzymes that covalently modify conserved lysine and serine residues. Both the phosphorylation of serine residues and the acetylation of lysine residues are associated with the modulation of transcriptional activity (see later). Ubiquitinylation of the C-terminal tail of H2A is also correlated with transcriptional activation. These modifications can be targeted through unknown mechanisms to particular chromosomal domains.

Several core histone variants exist that have specific alterations from normal histones in both the N-terminal tails and the DNA-binding surfaces of the histone-fold domains. For example, differences in amino acid sequence from the normal somatic H2A are conserved in a particular H2A variant (H2A.Z) from ciliate protozoa (*Tetrahymena*) to humans, and the histone H2A.Z variant in *Drosophila* has been shown to be essential for early development.

The core histones appear to have evolved from a DNA-binding protein that contained only the three α-helices of the histone-fold domain, lacking any tail domains. The archaebacterial protein, HMf, consists of only the histone-fold domain and has the capacity to wrap DNA around itself within nucleosome-like structures. The eukaryotic core histones have retained this property, but added the capacity of the assembled nucleoprotein complex to interact with other proteins outside the nucleosome through their additional tail domains.

The histone-fold is also found in some transcriptional regulatory proteins including $TAF_{II}40$ and $TAF_{II}60$ (TATA-box-binding protein associated factors) and the related CCAAT-box-binding proteins, NF-Y/CBF and HAP2/3/5. These use the histone-fold both to direct protein–protein interactions and to bind to DNA. $TAF_{II}40$ and $TAF_{II}60$ exist as a heterodimer in the general transcription factor TFIID. $TF_{II}40$ resembles H3 and $TAF_{II}60$ resembles H4. Both proteins have extended C-terminal tails that interact with other components of TFIID and transcriptional activators. It has been proposed that $TAF_{II}40$ and $TAF_{II}60$ participate in the assembly of nucleosome-like structures, excluding normal histones from the TATA box, yet maintaining DNA in a semi-compacted state competent for transcription [4]. Metazoan NF-Y/CBF and *Saccharomyces cerevisiae* HAP2/3/5 are highly related trimeric transcriptional activators. The evolutionarily conserved peptide sequences of two of the subunits (CBF-A and CBF-C or HAP3 and HAP5) resemble the histone-fold domains of histones H2B and H2A, respectively. These domains are essential for DNA binding in the presence of the third protein (CBF-B or HAP2) that confers sequence specificity.

Deviant nucleosomes — the 'winged helix' connection

Linker histones such as histone H1 or H5 contain a structured nucleic acid-binding domain known as the 'winged helix'. This domain is found in a variety of sequence-specific transcriptional regulators, including the prokaryotic catabolite gene activator protein (CAP) and the eukaryotic hepatocyte nuclear factor 3 (HNF3) protein. The winged helix consists of a bundle of three α-helices attached to a three-stranded anti-parallel β-sheet [5]. HNF3 binds within the major groove of DNA via one of the α-helices, suggesting that the structured domain of the linker histone will contact nucleosomal DNA in the same way (Figure 3). The linker histone contains additional basic N- and C-terminal domains that influence the path of linker DNA between adjacent nucleosomes.

Linker histones, such as histone H1, have been shown to direct the exact positioning of nucleosomes with respect to DNA sequence [6]. This positioning relies on the sequence- and structure-selective recognition of DNA by the linker histone, and protein–protein contacts made between the winged helix domain of the linker histone and the histone-fold domains of the core histones. During *Xenopus laevis* development, early embryonic variants of linker histones are replaced progressively with normal somatic histone H1. The inclusion of histone H1 in nucleosomes directs the positioning of the histone–DNA contacts over key promoter elements and represses transcription [7]. These transitions in chromatin composition and organization contribute directly to establishing the body plan of the embryo.

The mouse serum albumin enhancer exists in the active state within an array of precisely positioned nucleosome-like particles. Specific enhancer-binding factors, including HNF3, are part of the nucleosome-like particles, and HNF3 can actively direct their positioning with respect to DNA sequence [8]. These observations led to the hypothesis that HNF3 replaces linker histones within chromatin containing the serum albumin enhancer, thereby establishing a precise regulatory nucleoprotein architecture (Figure 4).

Transcription acetyltransferases

The synthesis of mRNA requires co-ordination of the activity of many transcription factors and enzymes. Specificity in transcriptional control relies upon the combinatorial binding of sequence-selective transcription factors to regulatory elements flanking the transcription start site (Figure 1). Activation domains within these factors recruit transcriptional co-activators that in turn enhance the activity of the basal transcriptional machinery. The role of co-activators is to integrate the signals from the various sequence-selective factors so that a final level of transcription can be determined. The simplest mechanism by which this might be achieved is for the co-activators to act as a scaffold between sequence-selective factors and the basal machinery [9].

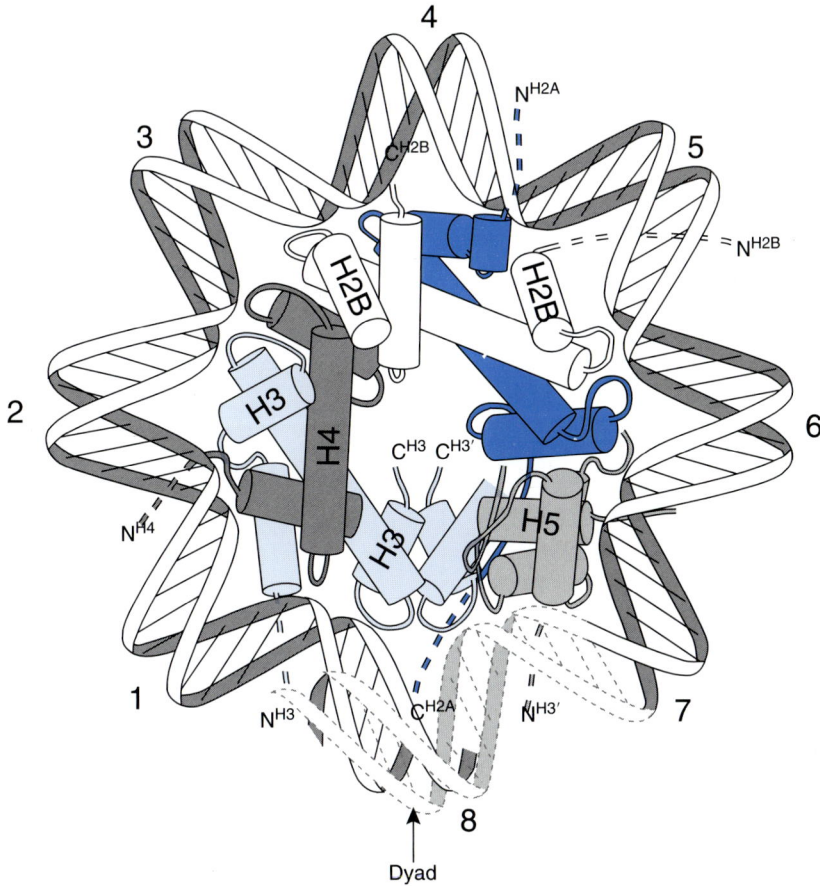

Figure 3. Model for the interaction of the histones with DNA in the nucleosome
This view is of one turn of DNA. For clarity only one molecule each of H2A, H2B and H4 is shown. Two molecules of H3 are shown to indicate the interface between the two (H3, H4) heterodimers. The numbers round the periphery of the diagram show turns of DNA away from the dyad axis. The structured domain of the linker histone (H5) is shown associating with nucleosomal DNA inside the super-helical turns of DNA around the histone octamer. The C- and N-termini of specific histones are shown by C^{H2B}, N^{H2B}, etc.

However, many co-activators also function as histone acetyltransferases. This introduces an alternative and potentially more dynamic model.

The Gcn5p–Ada2p–Ada3p complex in yeast is the archetypical transcription acetyltransferase and is a transcriptional co-activator targeted by transcription factors with acidic activation domains. The Gcn5p component of the complex has the capacity to acetylate specific lysines in histones H3 and H4 which are known to be associated with transcriptional activity [10]. Similar enzymic activities have now been found for P/CAF (a human homologue of Gcn5p), which associates with the p300/CBP [cAMP-response-element-binding protein (CREB)-binding protein] co-activator [11], and for p300/CBP

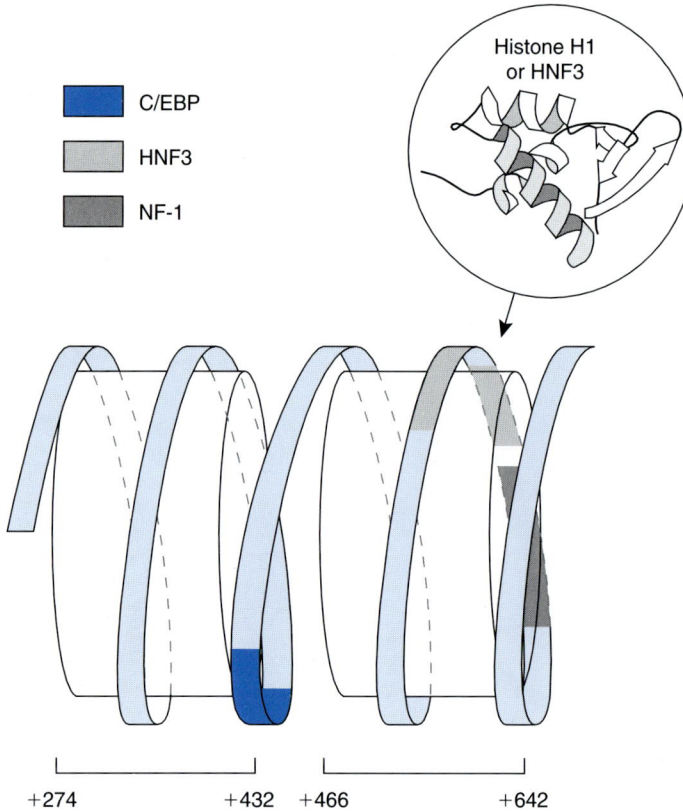

Figure 4. A specialized nucleosome on the mouse serum albumin enhancer
Two nucleosomes are shown positioned on the enhancer (numbers are relative to the 5′ end).
The boundaries of micrococcal nuclease digestion are indicated by the brackets. The positions of
transcription factor-binding sites are shown, as is the potential site of HNF3 or linker histone H1
interaction with the nucleosomal structures. The helix that interacts with DNA is shaded. C/EBP,
(CAAT/enhancer-binding protein; HNF3, hepatocyte nuclear factor3; NF-1, nuclear factor-1.
Reproduced from Wolffe, A.P. & Pruss, D. Deviant nucleosomes: the functional specialization of
chromatin. *Trends Genet.* **12**, 58–62 © (1996), with permission from Elsevier Science.

itself [12]. p300/CBP interacts with a variety of sequence-selective DNA-binding transcription factors, including nuclear hormone receptors, c-Jun/v-Jun,
c-Myb/v-Myb, c-Fos and MyoD. A core component of TFIID TAF$_{II}$250 also
has histone acetyltransferase activity [13].

Since core histone acetylation greatly facilitates the access of transcription
factors to the DNA in a nucleosome [14,15], and transcriptional co-activators
are histone acetyltransferases, a model can be proposed for transcriptional regulation in which histone acetylation directs the local destabilization of repressive histone–DNA interactions (Figure 1). Targeted acetylation allows the
basal transcriptional machinery to displace nucleosomes, assemble a functional
transcription complex and never have to deal with chromatin again. However,
a more interesting possibility follows from the discovery that transcriptional

regulators that deacetylate the histones exist [16]. This provides a molecular mechanism whereby transcription might be controlled continually. Core histones remain associated with DNA in the vicinity of a promoter in spite of the recruitment of the basal transcriptional machinery [17]. Thus the targeted or general activity of histone deacetylases will tend to return nucleosomes to their repressive configuration. The maintenance of gene activity would therefore require the continued activity of the co-activators as acetyltransferases. In this way transcriptional activity could be modulated continually through variation in chromatin conformation (Figure 5). These observations further emphasize that the eukaryotic transcriptional machinery is not only adapted to function in a chromatin environment, but also has the potential to make use of the packaging of DNA to regulate genes.

The SWI/SNF2 superfamily

Within *S. cerevisiae* the outcome of the interplay of transcription factors and histones at specific sites within gene promoters is influenced by the products of the *SWI1/ADR6*, *SWI/SNF2*, *SWI3*, *SNF5*, and *SNF6* genes [18]. All five of these proteins are found within a single 'general activator' complex required for the transcriptional induction of many yeast genes [18]. Genetic and biochemical studies of these yeast proteins, and their larger eukaryotic homologues, suggest that the general activator complex serves as a molecular machine to help transcription factors overcome the specific repressive effects of nucleosome assembly on transcription.

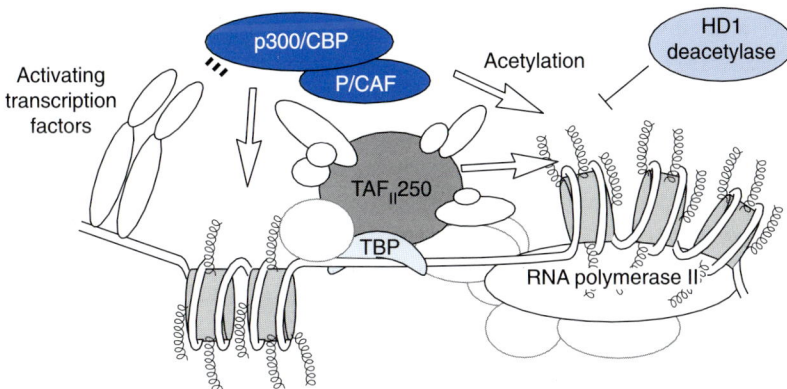

Figure 5. Transcriptional regulation in chromatin
Hormone-bound thyroid hormone receptor recruits a co-activator complex (p300/CBP–P/CAF) that retains chromatin in an 'open' configuration and a functional transcriptional machinery associated with the promoter. This complex counteracts the continued activity of the histone deacetylase (HD1). Reproduced from Wade, A.P., Pruss, D. & Wolffe, A. Histone acetylation: chromatin in action. *Trends Biochem. Sci.* **22**, 128–132 © (1997), with permission from Elsevier Science.

A major clue into the molecular mechanism by which the general activator complex exerts its function came from a genetic screen for mutations of genes that would allow transcription of the *HO* endonuclease gene (involved in yeast mating type switching) in the absence of SWI1 [19]. Two genes, *SIN1* and *SIN2*, were identified which, when mutated, led to SWI-independent transcription. Both of the *SIN* genes isolated in this way encode components of chromatin. SIN1 is a highly charged nuclear protein similar to mammalian high-mobility group proteins 1 and 2. A more direct association with nucleosomal structure is found for SIN2, which encodes histone H3 or H4 [19]. The SIN mutants in histones H3 and H4 cluster in one β-bridge motif within the heterodimer of H3 and H4. Due to the juxtaposition of two (H3, H4) heterodimers at the dyad axis of the nucleosome, the SIN mutations have the potential to disrupt histone–DNA interactions involving the central turn of DNA at the dyad axis (Figure 3). This could have a major impact on the integrity of both the nucleosome and higher-order chromatin structures.

The SWI2/SNF2 subunit of the general activator complex has a potent DNA-dependent ATPase activity [20]. Similar properties are associated with some DNA helicases; however, the general activator complex does not possess helicase activity, but may retain the capacity to track along DNA without unwinding the double helix. Such a processive movement may disrupt histone–DNA interactions. It is also possible that the general activator complex might mediate the removal of SIN1 or facilitate the displacement of histones (H2A, H2B) or (H3, H4)$_2$ through direct protein–protein or protein–DNA interactions. These will be progressively more difficult, as the stability of their interaction with DNA increases with their position towards the centre of the nucleosome. Biochemical analysis of chromatin following the activation of transcription by a mammalian steroid receptor, in a general-activator-dependent pathway, reveals only depletion of histone H1 [21]. Moreover any disruption of chromatin is rapidly reversible [22]. These results argue strongly against complete displacement of the core histones from DNA. The potential weakening of interaction between histones (H3, H4)$_2$ and DNA in the SIN2 mutants may facilitate local sliding of the entire histone octamer relative to regulatory elements, thereby facilitating access of the basal transcriptional machinery without octamer displacement [23]. SIN1 may interact with linker DNA in yeast, so removal of SIN1 might facilitate a comparable increase in histone octamer mobility and *trans*-factor access.

A major problem for the SWI/SNF family of ATPases has been the lack of proven mechanisms that might target their activities to known genes. However, it has now been shown that the association of the *Brahma*-related gene 1/BRG1-associated factor (BRG1–BAF) complex (a mammalian SWI/SNF homologue) with chromatin relies on the ligand-responsive glucocorticoid receptor [24]. The BRG1–BAF complex was known to facilitate transcriptional activation by ligand-bound glucocorticoid receptor, yet the molecular mechanisms by which this would be targeted were unclear. The

BRG1–BAF complex will disrupt nucleosomes *in vitro* and facilitate the binding of transcription factors to their recognition elements within chromatin, independent of any targeted transactivation domains. The recruitment of the BRG1–BAF complex to the glucocorticoid receptor *in vivo* is dependent upon a functional ligand-binding transactivation domain in the receptor. This result provides a direct connection between the *in vitro* chromatin-remodelling activities of the BRG1–BAF complex and the well established restructuring of nucleosomes on the glucocorticoid-responsive mouse mammary tumour virus chromatin used in these studies. It is presently unclear whether specific components of the BRG1–BAF complex make direct contact with the glucocorticoid receptor or if the association is indirect.

Additional support for the targeted association of SWI/SNF family ATPases comes from experiments in *Drosophila* and *Xenopus*. A distinct member of the SWI/SNF family, dMi-2, contributes to the determination of segmental identity during *Drosophila* development [25] by interacting with the Hunchback protein, which in turn binds directly to the regulatory elements of homoeotic genes to repress transcription. Biochemical experiments in *Xenopus* demonstrate that Mi-2 is part of a protein complex containing histone deacetylase [26]. Thus repression of homoeotic genes may involve histone deacetylation.

Future prospects

Histone acetyltransferases and deacetylases are now implicated in the fundamental mechanisms of transcription control. In many instances the proteins with these activities had already been characterized as having important regulatory functions. A focus for current research is to determine the exact consequence of histone acetylation for these specific regulatory functions. It is also important to recognize that these functions are likely to reveal a close relationship with the role in histone acetylation. Our knowledge of these issues is still far from complete; nevertheless the study of regulated histone acetylation has opened a window for visualizing chromatin in action. With respect to the SWI/SNF complex, unsolved questions include: determining exactly how the targeting of the complex to a specific promoter is directed, the nature of the molecular mechanisms directing chromatin disruption and how the process of gene activation is reversed. Resolution of these issues will require considerable progress in determining both the structural and functional role of SIN1 within chromatin, and the structural consequences of the SIN2 mutations in histones H3 and H4.

The discovery of novel structural subunits and regulatable physical interactions between chromatin components and SWI/SNF ATPase family members promises to integrate this interesting family of enzymes into diverse signal transduction pathways. The genomes of yeast and *Caenorhabditis elegans* suggest that these organisms contain 17 and 21 family members, respectively [27]. Many of these proteins will be found in distinct regulatory complexes that may be involved in gene activation, repression or both. It is also undoubtedly true that SWI/SNF ATPases will contribute to chromatin and chromosomal dynamics associated with the many other nuclear events that use DNA as a template.

Summary

- *A wide variety of histone-like proteins can be assembled into nucleosomal structures.*
- *Core and linker histone variants, proteins of the histone-fold and winged-helix families can all contribute to the local differentiation of functional chromosomal domains.*
- *It is very difficult to disrupt core histone interactions within a nucleosome in vivo. Histones H3 and H4 do not exchange out of chromatin outside S-phase. Histones H2A and H2B do exchange out of chromatin, but do so predominantly during transcription. This confers stability on the nucleosome during the cell cycle. Linker histones have a much less stable association with nucleosomal DNA, allowing for reversible activation of transcription.*
- *A distinct feature of histone interactions with nucleosomal DNA is the exposure of DNA on the surface of the nucleosome. One side of DNA is occluded on the histone surface, but the other is exposed and potentially accessible to other regulatory proteins.*
- *A major contributory factor to the functional specialization of chromatin is the capacity to target nucleosome modification and disruption.*

References

1.   Schild, C., Claret, F.-X., Wahli, W. & Wolffe, A.P. (1993) A nucleosome-dependent static loop potentiates estrogen-regulated transcription from the *Xenopus* vitellogenin B1 promoter *in vitro*. *EMBO J.* **12**, 423–433
2.   Arents, G., Burlingame, R.W., Wang, B.W., Love, W.E. & Moudrianakis, E.N. (1991) The nucleosomal core histone octamer at 3.1Å resolution: a tripartite protein assembly and a left-handed superhelix. *Proc. Natl. Acad. Sci. U.S.A.* **88**, 10148–10152
3.   Luger, K., Mader, A.W., Richmond, R.K., Sargent, D.F. & Richmond, T.J. (1997) Crystal structure of the nucleosome core particle at 2.8Å resolution. *Nature (London)* **389**, 251–260
4.   Hoffmann, A., Chiang, C.-M., Oelgeschlager, T., Xie, X., Burley, S.K., Nakatani, Y. & Roeder, R.G. (1996) A histone octamer-like structure within TFIID. *Nature (London)* **380**, 356–360
5.   Ramakrishnan, V., Finch, J.T., Graziano, V., Lee, P.L. & Sweet, R.M. (1993) Crystal structure of globular domain of histone H5 and its implications for nucleosomal binding. *Nature (London)* **362**, 219–223
6.   Sera, T. & Wolffe, A.P. (1998) The role of histone H1 as an architectural determinant of chromatin structure and as a 5S rRNA gene. *Mol. Cell. Biol.* **18**, 3668–3680
7.   Bouvet, P., Dimitrov, S. & Wolffe, A.P. (1994) Specific regulation of *Xenopus* chromosomal 5S rRNA gene transcription *in vivo* by histone H1. *Genes Dev.* **8**, 1147–1159

8. Cirillo, L.A., McPherson, C.E., Bossard, P., Stevens, K., Cherian, S., Shim, E.Y., Clark, K.L., Burley, S.K. & Zaret, K.S. (1998) Binding of the winged-helix transcription factor HNF3 to a linker histone site on the nucleosome. *EMBO J.* **17**, 244–254

9. Verrijzer, C.P. & Tjian, R. (1996) TAFs mediate transcriptional activation and promoter selectivity. *Trends Biochem. Sci.* **21**, 338–342

10. Zhang, W., Bone, J.R., Edmondson, D.G., Turner, B.M. & Roth, S.Y. (1998) Essential and redundant functions of histone acetylation revealed by mutation of target lysines and loss of the Gcn5p acetyltransferase. *EMBO J.* **17**, 3155–3167

11. Yang, X.-J., Ogryzko, V.V., Nishikawa, J.-I., Howard, B.H. & Nakatani, Y. (1996) A p300/CBP-associated factor that competes with the adenoviral E1A oncoprotein. *Nature (London)* **382**, 319–324

12. Ogryzko, V.V., Schiltz, R.L., Russanova, V., Howard, B.H. & Nakatani, Y. (1996) The transcriptional coactivators p300 and CBP are histone acetyltransferases. *Cell* **87**, 953–959

13. Mizzen, C.A., Yang, X.J., Kobuko, T., Brownell, J.E., Bannister, A.J., Owen-Hughes, T., Workman, J., Wang, L., Berger, S.L., Kouzarides, T., et al. (1996) The $TAF_{II}250$ subunit of TFIID has histone acetyltransferase activity. *Cell* **87**, 1261–1270

14. Lee, D.Y., Hayes, J.J., Pruss, D. & Wolffe, A.P. (1993) A positive role for histone acetylation in transcription factor binding to nucleosomal DNA. *Cell* **72**, 73–84

15. Vettesse-Dadey, M., Grant, P.A., Hebbes, T.R., Crane-Robinson, C., Allis, C.D. & Workman, J.L. (1996) Acetylation of histone H4 plays a primary role in enhancing transcription factor binding to nucleosomal DNA *in vitro*. *EMBO J.* **15**, 2508–2518

16. Taunton, J., Hassig, C.A. & Schreiber, S.L. (1996) A mammalian histone deacetylase related to a yeast transcriptional regulator Rpd3. *Science* **272**, 408–411

17. Nacheva, G.A., Guschin, D.Y., Preobrazhenskaya, O.V., Karpov, V.L., Ebralidse, K.K. & Mirzabekov, A.D. (1989) Change in the pattern of histone binding to DNA upon transcriptional activation. *Cell* **58**, 27–36

18. Peterson, C.L. & Tamkun, J.W. (1995) The SWI/SNF complex: a chromatin remodeling machine? *Trends Biochem. Sci.* **20**, 143–146

19. Kruger, W., Peterson, C.L., Sil, A., Coburn, C., Arents, G., Moudrianakis, E.N. & Herskowitz, I. (1995) Amino acid substitutions in the structured domains of histones H3 and H4 partially relieve the requirement of the yeast SWI/SNF complex for transcription. *Genes Dev.* **9**, 2770–2779

20. Laurent, B.C., Treioch, I. & Carlson, M. (1993) The yeast SNF2/SWI2 protein has DNA stimulated ATPase activity required for transcriptional activation. *Genes Dev.* **7**, 583–591

21. Bresnick E.H., Bustin, M., Marsaud, V., Richard-Foy, H. & Hager, G.L. (1992) The transcriptionally-active MMTV promoter is depleted of H1. *Nucleic Acids Res.* **20**, 273–278

22. Lee, H.H. & Archer, T.K. (1994) Nucleosome-mediated disruption of transcription factor-chromatin initiation complexes at the mouse mammary tumor virus long terminal repeat *in vivo*. *Mol. Cell. Biol.* **14**, 32–41

23. Wolffe, A.P. (1994) Switched-on chromatin. *Curr. Biol.* **4**, 525–527

24. Fryer, C.J. & Archer, T.K. (1998) Chromatin remodeling by the glucocorticoid receptor requires the BRG1 complex. *Nature (London)* **393**, 88–91

25. Kehle, J., Beuchle, D., Treuheit, S., Christen, B., Kennison, J.A., Bienz, M. & Muller, J. (1998) dMi2, a hunchback-interacting protein that functions in *Polycomb* repression. *Science* **282**, 1897–1900

26. Wade, P.A., Jones, P.L., Vermaak, D. & Wolffe, A.P. (1998) A multimple subunit histone deacetylase from Xenopus laevis contains a Snf2 superfamily ATPase. *Curr. Biol.* **8**, 843–846

27. Chervitz, S.A., Aravind, L., Sherlock, G., Ball, C.A., Koonin, E.V., Dwight, S.S., Harris, M.A., Dolinski, K., Mohr, S., Smith, T., et al. (1998) Comparison of the complete protein sets of worm and yeast: orthology and divergence. *Science* **282**, 2022–2028

5

DNA methylation and control of gene expression in vertebrate development

Richard R. Meehan[1] and Irina Stancheva

Genes and Development Group, University of Edinburgh, Deptartment of Biomedical Sciences, Hugh Robson Building, George Square, Edinburgh EH8 9XD, Scotland, U.K.

Introduction

DNA methylation has the ability to repress transcription; in addition, the pattern of methylation can be stably inherited during successive cell divisions [1]. These two features make methylation very attractive as a potential regulator of gene expression during the development of an organism. A combination of biochemical and genetic analysis has elucidated the potential mechanisms by which methylation can inhibit transcription, and has shown that DNA methylation is necessary for normal development in many vertebrates [2–4]. Indeed a broader view would indicate that DNA methylation has been successfully utilized in many eukaryotes, including fungi, plants and animals, as a regulator of gene activity [1]. What is not clear is whether the specific role of DNA methylation is conserved between these different species or whether methylation has been adapted to regulate different aspects of gene expression in diverse species. This chapter will attempt to address this question, principally by comparing the effects of depletion of DNA methylation on the development of vertebrates: mice (*Mus musculus*), zebrafish (*Danio rerio*) and the toad (*Xenopus laevis*).

[1]*To whom correspondence should be addressed (e-mail: Richard.Meehan@ed.ac.uk).*

DNA methylation in vertebrates

Vertebrate DNA is methylated at the fifth position of cytosine (^{5}mC) in the dinucleotide CpG. This is carried out enzymically by DNA methyl-transferases either immediately after passage of the replication fork or as a result of repair processes [1]. Approximately 70% of all methylated CpGs (^{Me}CpGs) are found in regions of the genome that are transcriptionally inactive and late replicating. An early immuno-histochemical study with an anti-^{5}mC antibody showed that mouse heterochromatin is rich in methylated sequences [5]. Interest in the role of DNA methylation in controlling gene expression was stimulated by the finding that many tissue-specific genes are methylated in non-expressing tissues [6]. However, the view that DNA methylation is necessary for gene inactivation was challenged when it became clear that there are many tissue-specific genes that are never methylated even in tissues where they are not expressed [7]. A general view is that DNA methylation acts as a global inhibitor of transcription due to the location of ^{5}mC throughout the genome, thus preventing background transcriptional noise or the spontaneous activation of normally silent chromosomal regions [7]. This does not rule out the possibility that specific sets of genes are regulated by DNA methylation at different points in development.

The importance of DNA methylation in mammalian development was emphasized when it was demonstrated that the targeted disruption of the maintenance DNA methyltransferase (*Dnmt1*) gene in mice resulted in embryonic lethality during early stages of development [2,8]. Several genes were mis-expressed in the *Dnmt1* $^{-/-}$ embryos [9], probably due to loss of ^{5}mC. However, tissue-specific genes were not activated in these mutants and, moreover, were not methylated during the early stages of development in wild-type mice [7]. This implies that the observed tissue-specific methylation patterns are a late developmental event and that DNA methylation may occur as a result of gene inactivation. It is not clear why the *Dnmt1*$^{-/-}$ mutants die, although a strong possibility is that it is due to the mis-expression of normally dormant methylated genes and the abnormal inactivation of the X chromosome(s) in male and female mice. Indeed it has been argued on this basis that DNA methylation has been adapted for a specialized role in mammalian development, controlling the expression of imprinted genes and X-inactivation, rather than being a general regulator of gene activity during embryogenesis [7,9]. One way to examine these possibilities is to compare the role of DNA methylation in different vertebrate species and to ask whether there are any similarities or differences. However, before we do that we should explore how DNA methylation can repress gene expression.

Gene repression by DNA methylation

It is very clear that methylation of genes *in vivo* and *in vitro* results in transcriptional silencing, the degree of which may depend upon the location and density of the CpGs relative to the promoter [10]. Several theories can account for how ^{Me}CpG can interfere with gene expression. First, methylation may directly prevent a transcription factor from recognizing its binding sites, because the ^{5}mC changes the recognition sequence [10]. However, many transcription factors (for example Sp1) bind independently of the methylation status of their recognition sequences and the recognition sequences, for many other transcription factors do not contain a CpG. Secondly, methylation may lead to the formation of a localized chromatin structure that is incompatible with gene expression [11,12]. Time course experiments have demonstrated that repression of transcription from methylated templates is not immediate. Methylated genes are remodelled over several hours into a structure that leads to their repression [13]. The preferential condensation of methylated chromatin into a higher-order structure is an attractive idea and has received support from an initial finding that methylated DNA is preferentially located in nucleosomes containing histone H1 [14]. However, with one notable exception, CpG methylation has little or no effect on the capacity of the histone octamer to interact with DNA [15]. Finally, it is possible that nuclear factors preferentially interact with methylated DNA leading to the formation of inactive chromatin (Figure 1). Such factors have been identified in many species and have been termed methylated-CpG-binding proteins (MeCPs) [10]. The next section discusses their identification and how their repressing activity is linked to chromatin-remodelling complexes.

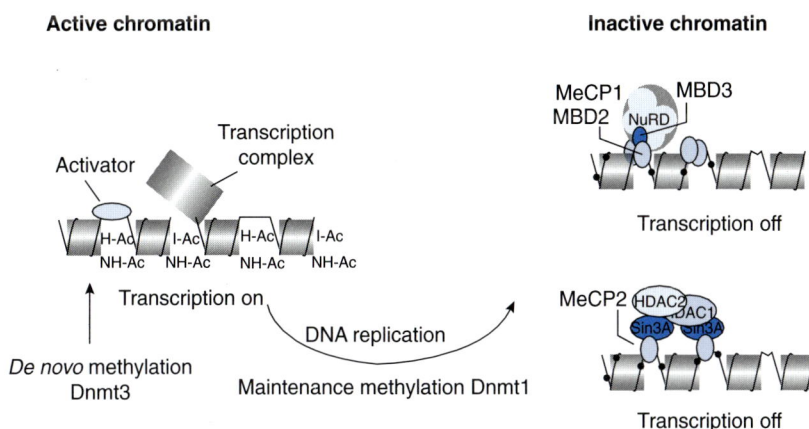

Figure 1. ^{Me}CpG-binding proteins (MeCP1, MeCP2, MBD1 and MBD2) can target chromatin-remodelling complexes to methylated sequences
These complexes (NuRD and Sin3A) contain histone deacetylase (HDAC) activities which promote the establishment of an inactive chromatin conformation by removing acetyl groups from acetylated histone tails (NH-Ac-). Black dots indicate ^{Me}CpGs which act as ligands for the MeCPs.

Methylated-DNA-binding proteins

The first factor that showed a binding preference for methylated DNA was identified in a filter-binding assay using human placental extracts [16]. Subsequently, band-shift and Southwestern assays were used to identify similar factors, termed MeCP1 and MeCP2, in rodent extracts [17–19]. Both factors are highly specific for symmetrically methylated CpG pairs. Hemi-methylated DNA or ^{5}mC in a non-CpG context were not substrates. When methylated templates were tested in F9 embryonal carcinoma cells, which contain low levels of MeCPs, they were found to be transcribed at high levels, whereas they were repressed in fibroblast cell lines which contained high MeCP activity. This implied that MeCPs could be transcriptional repressors. The MeCP2 protein was purified, sequenced and the corresponding cDNA isolated [19]. MeCP2 protein could be found throughout mouse chromosomes, but was especially concentrated in the heavily methylated centromeric heterochromatin. MeCP2 has a modular structure and contains a methylated DNA-binding domain (MBD) that is responsible for targeting it to mouse heterochromatin [20]. Screening of human and mouse databases indicated that there is a large family of proteins (MBD1–MBD4) that contain this motif [21,22]. With the exception of MBD3, each is capable of binding specifically to methylated DNA [19,20,23,24]. MeCP2, MBD1 and MBD2 were also found to repress transcription from methylated templates [21,24–27]. MBD4 has protein identity with bacterial DNA repair enzymes and is a glycosylase that can efficiently remove thymine or uracil from a mismatched CpG *in vitro* [28]. Although a truncated form of MBD2 has been reported to have a demethylase activity [29], other groups could not substantiate this observation [25,26]. It is more likely that MBD2 acts principally as a transcriptional repressor.

MBD-containing proteins repress transcription from methylated templates

How MeCP2 can repress transcription from methylated templates was initially investigated by the use of nuclear extracts that support transcription *in vitro*. Under conditions where both methylated and non-methylated templates are transcribed, it was demonstrated that the addition of recombinant MeCP2 inhibited the methylated template preferentially [27]. The MBD domain of MeCP2 was required and, through the use of GAL4 fusions and a GAL4-dependent template, it was demonstrated that there is a separate transcription repression domain. In these experiments only naked DNA was used, but it could be demonstrated that recombinant MeCP2 was incorporated preferentially into methylated chromatin, displacing histone H1 in the process [27]. Subsequently it has been shown that *Xenopus* MeCP2 can associate with ^{Me}CpGs in a mononucleosome [30]. Core histones can also be modified (by methylation, phosphorylation and acetylation) and this can have functional consequences [31]. There is a general correlation between acetylation of the N-terminal tails of the core histones and a more open chromatin structure that facilitates gene expression (Figure 1). Many transcriptional co-activators have histone acetyltransferase activity and, conversely, there are transcriptional co-repressors that have histone deacetylase activity. MeCP2 was co-purified with a histone deacetylase complex, and inhibitors of histone deacetylase relieve MeCP2-mediated repression [24,25]. In HeLa cells, MBD2 is associated with a histone deacetylase in the MeCP1 repressor complex, and MBD3 has been found in a separate Mi2/NuRD deacetylase complex that posses a nucleosome-remodelling activity [32]. These observations provide a direct link between the methylated DNA and chromatin configuration. Future work will show how distinct these complexes are and determine whether there is cross-talk between different MeCpG-binding proteins within the biochemically distinct histone deacetylase complexes. It is possible that MeCP2/MBDs target these modifying complexes to distinct regions of the genome.

DNA methyltransferases

Four mammalian cytosine DNA methyltransferases (Dnmt) have been identified [1], all of which contain a highly conserved C-terminal catalytic domain and variable N-terminal extensions (Figure 2). With the exception of Dnmt2, they specifically methylate cytosine in CpG. Dnmt1 is the best studied and has a 500-residue C-terminal catalytic domain and a 1100-residue N-terminal extension that has a regulatory role with respect to substrate specificity and in targeting the methyltransferase to different nuclear and

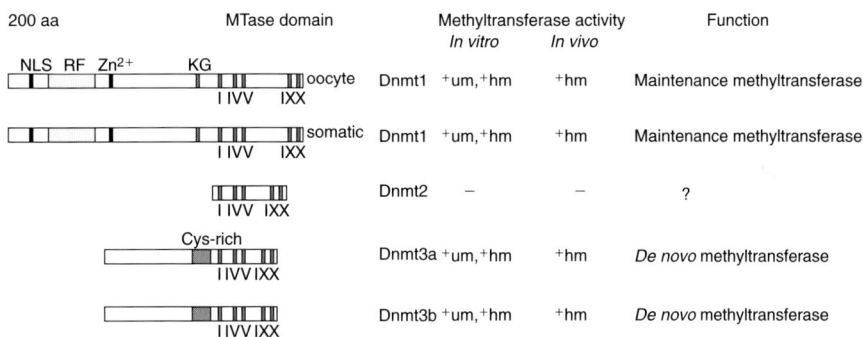

Figure 2. The family of vertebrate DNA methyltransferase enzymes
The conserved regions in the methyltransferase catalytic domains are indicated with black boxes (I, IV, V, IX and X). In Dnmt1, NLS is a nuclear localization region, RF is the region of the protein targeting the enzyme to replicating foci and Zn^{2+} is a zinc-binding motif. KG, lysine/glycine repeats. Both Dnmt3a and Dnmt3b have a cysteine-rich region that may bind DNA. hm, hemi-methylated DNA; um, non-methylated DNA. + and − indicate the level of activity towards its substrate. Note that Dnmt1 has different activities *in vitro* and *in vivo*.

cellular sites (Figure 2). Hemi-methylated DNA is the preferred substrate for Dnmt1, but *in vitro* it possesses some activity towards non-methylated DNA (*de novo* activity). Since genomic DNA is in the hemi-methylated state after replication, Dnmt1 is regarded as the maintenance methylase, propagating pre-existing patterns of methylation. Recent additions to the vertebrate cytosine methyltransferase family are Dnmt3a and Dnmt3b [33], which were identified by screening mouse and human databases with sequences corresponding to the catalytic domain. These enzymes have *de novo* methyltransferase activity *in vitro* and *in vivo* and are required at different periods of mouse development [34].

Patterns of DNA methylation during vertebrate development

A big question is 'How are DNA methyltransferases recruited to, or excluded from, particular regions of the genome during development?' In mice there are complex changes in DNA methylation levels during embryogenesis and cell differentiation [35]. This involves a global demethylation during cleavage followed by a wave of *de novo* methylation in the growing embryo (Figure 3). These alterations in methylation appear to be intimately associated with the phenomenon of imprinting in mammals, whereby the expression pattern of a gene can be influenced by whether it is paternally or maternally inherited. The establishment of imprinted epigenetic marks takes place during gametogenesis, a process that is related to differential methylation of paternal and maternal alleles of some genes that are essential for embryo growth [9]. Initially the germ line cells are undermethylated. In fully mature gametes, sperm DNA has higher levels of methylation than oocyte DNA. Immunohistochemical experiments with a monoclonal antibody against ^{5}mC , and bisulphate sequencing of maternal and paternal alleles of imprinted genes, have demonstrated that the male pronucleus is rapidly demethylated after fertilization by an activity present in the oocyte cytoplasm [36]. The maternal genome remains more methylated than the paternal DNA up to the 8-cell stage, when they become equivalent. The level of methylation decreases to 15% in the early blastocysts and returns to higher levels during implantation. Meanwhile the maternal oocyte form of Dnmt1 is excluded from the nucleus, and its relocation there is coincident with the remethylation of the mouse genome [37]. It has been argued that there is a highly regulated process of reprogramming in the developing mouse embryo that involves erasing of epigenetic modifications present in the zygote followed by subsequent *de novo* methylation necessary for resetting the developmental patterns of gene expression in differentiating cell lineages [35].

In contrast to mammals, neither *X. laevis* nor zebrafish have imprinted genes or identifiable sex chromosomes. Compared with the dynamic changes of ^{5}mC levels in mammals, the early embryonic patterns of methylation in

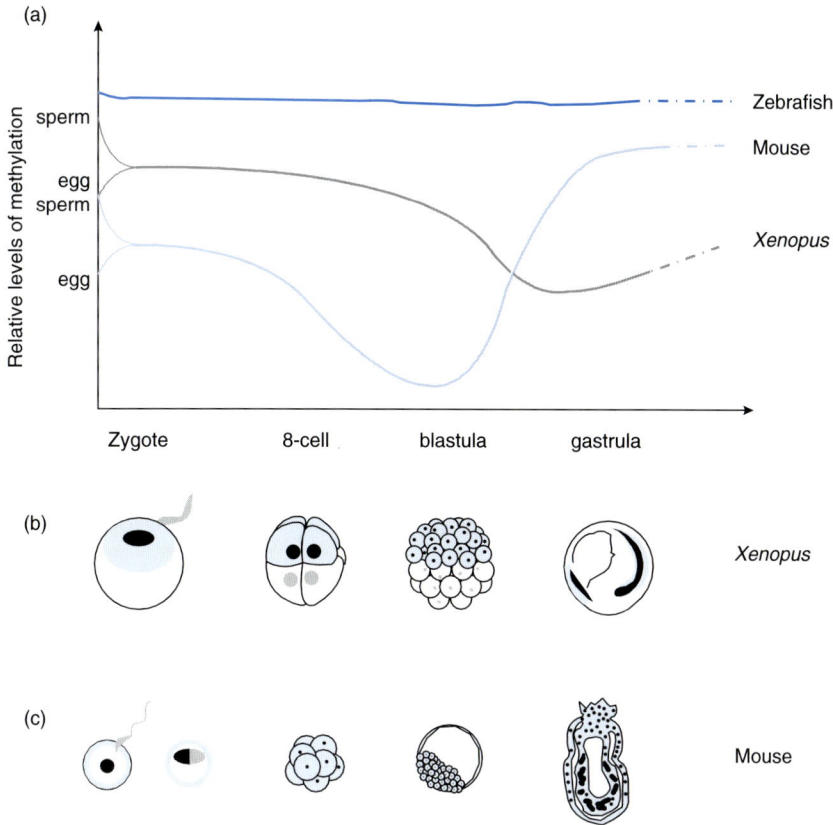

Figure 3. Relative levels of methylation and localization of Dnmt1 enzyme during mouse, *Xenopus* and zebrafish development

(a) Compared with the dynamic alterations in ^{5}mC content during early mouse development, methylation changes are much less pronounced in amphibia and absent in fish. The methylation patterns of mammalian somatic cell lineages are established during gastrulation. In lower vertebrates, early embryonic cells are partially fated to differentiate before gastrulation. (b) In *X. laevis* and zebrafish, Dnmt1 (blue) is asymmetrically localized in the oocyte cytoplasm and blastula stage embryos. Consequentially, DNA (black) of *Xenopus* animal pole cells is more methylated than that of the vegetal hemisphere cells (grey). (c) In mouse, sperm DNA (black) is actively demethylated in the oocyte cytoplasm after fertilization, while the female pronucleus remains unchanged. The ^{5}mC content of the zygote decreases progressively and is very low in the blastocyst. Somatic methylation patterns are established in the gastrula. Dnmt1 (blue) migrates from cytoplasm to the nucleus at the 8-cell stage, but is excluded later during the demethylation of blastocyst cells.

these animals are set up in a different manner. Dnmt1 is asymmetrically localized in the oocyte and egg cytoplasm of toads and fish [3,4]. The animal pole of the egg, which gives rise to mesoderm and ectodermal tissues, contains high levels of Dnmt1 RNA and protein, while the vegetal hemisphere, that later differentiates into ectoderm, contains considerably lower levels of the enzyme. Such localization perhaps has a functional relevance, since up to 3-fold differences in the amount of ^{5}mC can be detected around midblastula in DNA derived from *Xenopus* animal and vegetal cells. As in the mouse, DNA methylation decreases during the cleavages, but overall changes in ^{5}mC content are not so dramatic [4]. Low levels of methylation coincide with the initiation of zygotic transcription at midblastula and, in contrast to mouse, remain relatively low during gastrulation. At the other extreme, there are no detectable changes in methylation levels during zebrafish development [38]. Most likely, amphibia and fish do not need a genome-wide wave of demethylation and remethylation to reset the initial methylation patterns, since they might be established already in sub-populations of cells before gastrulation. This observation is supported by the finding that *Xenopus* blastomeres are already fated to differentiate to different lineages in 8-cell and 16-cell embryos [39], while embryonic stem cells derived from mouse blastocysts are uniformly methylated and totipotent. Methylation patterns of differentiated somatic cells in lower vertebrates and mammals are also probably achieved by tissue-specific expression of *de novo* and maintenance methyltransferase enzymes.

DNA methylation is essential for vertebrate development

In all species studied to date (mouse, zebrafish and *Xenopus*), Dnmt1 is expressed as an abundant oocyte form and later shows varying tissue-specific levels [3,4,37]. Loss of Dnmt1 or inhibition by 5-azacytidine (an analogue of cytidine) are injurious to embryo development in all three species [2–4]. Due to the phenotypic complexity and early lethality of $Dnmt1^{-/-}$ mutants in mice, the question as to whether methylation is essential for regulation of gene expression at the onset of gastrulation in the developing embryo has been difficult to answer [2,8]. Studies have shown that loss of methylation affects mesoderm formation during gastrulation [3,4] of Dnmt1-deficient *Xenopus* and zebrafish embryos grown in the presence of 5-azacytidine. $Dnmt1^{-/-}$ mice exhibit development delay and asynchrony, which may also be indicative of gastrulation abnormalities [2,8]. In *Xenopus* DNA, methylation is involved in the maintenance of genome-wide transcriptional silencing that occurs between fertilization and midblastula transition [4]. It is also important in the regulation of appropriate expression patterns of developmentally essential genes, among which are transcription factors (*Xbra* and *Otx2*) and signalling molecules (*Cerberus*). Transient depletion of maternally expressed Dnmt1 by anti-sense RNA in *Xenopus* leads to premature transcription of these genes at least two cell cycles earlier than normal. Loss of the initial methylation patterns and inappropriate gene activation lead to changes in the developmental competence of animal pole cells. Similar defects were observed in zebrafish embryos grown in the presense of 5-azacytidine. One generalization from the analysis of Dnmt1 depletion or inhibition in all three species is that it is not essential for the survival of embryonic cells during early cleavage stages. The effect of disrupting pre-existing methylation patterns only becomes apparent during and after gastrulation. The common features of methylation-depleted embryos in all three species include a failure to organize

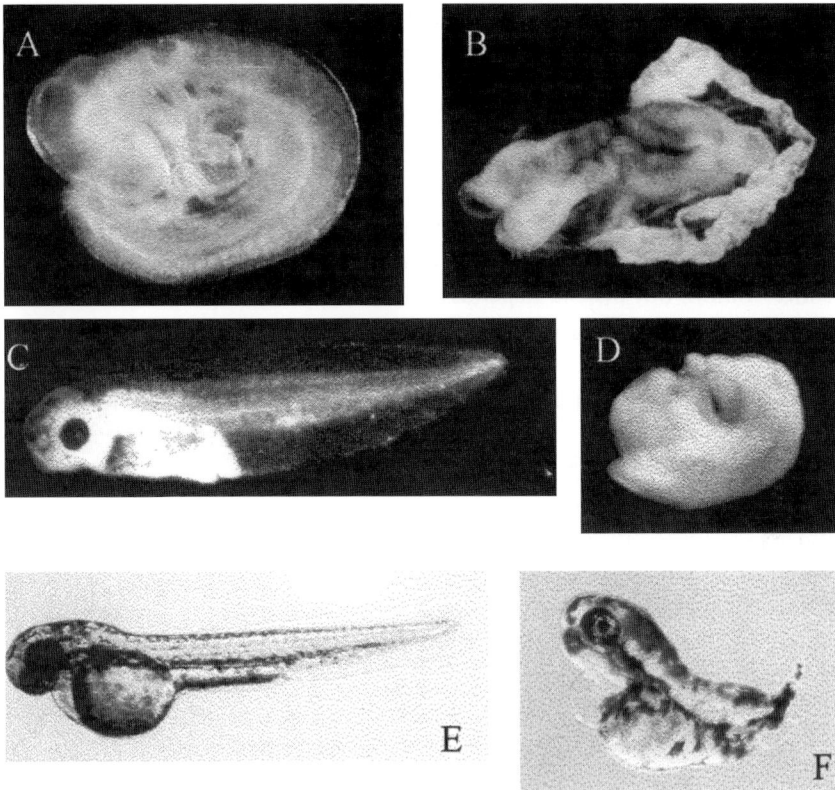

Figure 4. Loss or inhibition of Dnmt1 results in severe developmental defects
(a) Normal mouse embryo at embryonic stage 14. (b) $Dnmt1^{-/-}$ mutant mouse embryos fail to develop and die at about stage 14. (c) Normal *Xenopus* tadpole (stage 35). (d) *Xenopus* embryo depleted of Dnmt1 during cleavage stages. These embryos do not form head and axial structures (equivalent of stage 35). (e) Normal zebrafish embryo after 48 h of development. (f) About 30% of zebrafish embryos treated with 5-azacytidine show reduction of the dorsal axis and abnormal patterning of the somites. Reproduced, with permission, from *Dev. Biol.* **206**, 189–205 © (1999) Academic Press, Inc., and *Development* **122**, 3195–3205 © (1996) Company of Biologists

neural and muscle tissues and a high degree of apoptosis (Figure 4). The accumulated evidence argues that methylation-mediated transcriptional repression is important for the normal development of all vertebrates.

Future perspectives

Nuclear transplantation experiments in mammals and amphibia have shown that, in rare cases, somatic nuclei are able to support normal embryonic development [40,41]. Presumably, the somatic patterns of methylation have to be erased to allow the full potential of the transferred nucleus to be realized. This suggests that epigenetic mechanisms may have a role in potentiating embryonic cells to differentiate by reinforcing, through DNA methylation and the modification of chromatin structure, particular expression states. The recent findings that methyl-binding proteins are involved in chromatin remodelling by targeting of large protein complexes that involve Sin3A/histone deacetylase complex or NuRD to methylated DNA will ultimately answer the question as to how the expression of specific genes is regulated by methylation [31].

From a developmental point of view it is essential to know how DNA methylation patterns and the subsequent gene repression are established. It is tempting to speculate that DNA methylation is not simply a consequence of gene inactivation, but instead depends upon active targeting of *de novo* DNA methyltransferases to specific sequences. It is also possible that there is developmentally controlled feedback between the maintenance and *de novo* methyltransferases in the establishment of differential tissue-specific methylation patterns.

Another question is whether DNA methylation in vertebrates is involved in the regulation of the similar sets of genes (apart from the imprinted loci in mammals) during embryogenesis. Unfortunately this question has not been answered in mice because of the complexity of Dnmt1 phenotypes, which are greatly dominated by the negative effects of mis-expression of imprinted genes and X-chromosome inactivation. Since most of the components of the methylation-mediated gene repression machinery are well conserved between *Xenopus*, mouse and human, this raises the possibility that methylation plays a very similar function in the development of all vertebrates.

Summary

- *^{Me}CpGs act as ligands for nuclear factors (repressors) that are components of chromatin modification and remodelling activities.*
- *The DNA-methylation-mediated repression system (Dnmt1s, MeCPs and MBDs) is highly conserved in vertebrates.*
- *DNA methylation is essential for normal vertebrate development.*
- *It is possible (but remains unproven) that the role of DNA methylation in regulating development is highly conserved in vertebrates.*
- *In mammals, DNA methylation has an additional role in regulating the expression of imprinted genes and in controlling X-inactivation.*

References

1.   Colot, V. & Rossignol, J.-L. (1999) Eukaryotic DNA methylation as an evolutionary device. *BioEssays* **21**, 402–411
2.   Li, E., Bestor, T.H. & Jaenisch, R. (1992) Targeted mutation of the DNA methyltransferase gene results in embryonic lethality. *Cell* **69**, 915–926
3.   Martin, C.C., Laforest, L., Akimenko, M.A. & Ekker, M. (1999) A role for DNA methylation in gastrulation and somite patterning. *Dev. Biol.* **206**, 189–205
4.   Stancheva, I. & Meehan, R.R. (2000) Transient depletion of xDnmt1 leads to premature gene activation in *Xenopus* embryos. *Genes Dev.* **14**, 313–327

5.   Miller, O., Schnedl, W., Allen, J. & Erlanger, B. (1974) 5-Methylcytosine localised in mammalian
     constitutive heterochromatin. *Nature (London)* **251**, 636–637

6.   Jones, P.A. (1998) DNA methylation paradox. *Trends Genet.* **15**, 34–37

7.   Walsh, C.P. & Bestor, T.H. (1999) Cytosine methylation and mammalian development. *Genes Dev.*
     13, 26–34

8.   Lei, H., Oh, S.P., Okano, M., Juttermann, R., Goss, K.A., Jaenisch, R. & Li, E. (1996) De novo DNA
     cytosine methyltransferase activities in mouse embryonic stem cells. *Development* **122**,
     3195–3205

9.   Jaenisch, R. (1997) DNA methylation and imprinting: why bother? *Trends Genet.* **13**, 323–329

10. Bird, A. (1992) The essentials of DNA methylation. *Cell* **70**, 5–8

11. Kass, S.U., Goddard, J.P. & Adams, R.L. (1993) Inactive chromatin spreads from a focus of methyl-
     ation. *Mol. Cell. Biol.* **13**, 7372–7379

12. Tate, P.H. & Bird, A.P. (1993) Effects of DNA methylation on DNA-binding proteins and gene
     expression. *Curr. Opin. Genet. Dev.* **3**, 226–231

13. Kass, S.U., Landsberger, N. & Wolffe, A.P. (1997) DNA methylation directs a time-dependent
     repression of transcription initiation. *Curr. Biol.* **7**, 157–165

14. Ball, D.J., Gross, D.S. & Garrard, W.T. (1983) 5-Methylcytosine is localized in nucleosomes that
     contain histone H1. *Proc. Natl. Acad. Sci. U.S.A.* **80**, 5490–5494

15. Davey, C., Pennings, S. & Allan, J. (1997) CpG methylation remodels chromatin structure *in vitro*.
     J. Mol. Biol. **267**, 276–288

16. Huang, L.-K., Wang, R., Gama-Sosa, M.A., Shenoy, S. & Ehrlich, M. (1984) A protein from human
     placental nuclei binds preferentially to 5-methylcytosine-rich DNA. *Nature (London)* **308**, 293–295

17. Meehan, R.R., Lewis, J.D., McKay, S., Kleiner, E.L. & Bird, A.P. (1989) Identification of a mam-
     malian protein that binds specifically to DNA containing methylated CpGs. *Cell* **58**, 499–507

18. Meehan, R.R., Lewis, J.D. & Bird, A. (1992) Characterization of MeCP2, a vertebrate DNA binding
     protein with affinity for methylated DNA. *Nucleic Acids Res.* **20**, 5085–5092

19. Lewis, J.D., Meehan, R.R., Henzel, W.J., Maurer-Fogy, I., Jeppesen, P., Klein, F. & Bird, A. (1992)
     Purification, sequence, and cellular localization of a novel chromosomal protein that binds to
     methylated DNA. *Cell* **69**, 905–914

20. Nan, X., Tate, P., Li, E. & Bird, A. (1996) DNA methylation specifies chromosomal localization of
     MeCP2. *Mol. Cell. Biol.* **16**, 414–421

21. Cross, S.H., Meehan, R.R., Nan, X. & Bird, A. (1997) A component of the transcriptional repres-
     sor MeCP1 shares a motif with DNA methyltransferase and HRX proteins. *Nat. Genet.* **16**,
     256–259

22. Hendrich, B. & Bird, A. (1998) Identification and characterization of a family of mammalian
     methyl-CpG binding proteins. *Mol. Cell. Biol.* **18**, 6538–6547

23. Wade, P.A., Gegonne, A., Jones, P.L., Ballestar, E., Aubry, F. & Wolffe, A.P. (1999) Mi-2 complex
     couples DNA methylation to chromatin remodelling and histone deacetylation. *Nat. Genet.* **23**,
     62–66

24. Nan, X., Ng, H.H., Johnson, C.A., Laherty, C.D., Turner, B.M., Eisenman, R.N. & Bird, A. (1998)
     Transcriptional repression by the methyl-CpG-binding protein MeCP2 involves a histone deacety-
     lase complex. *Nature (London)* **393**, 386–389

25. Jones, P.L., Veenstra, G.J., Wade, P.A., Vermaak, D., Kass, S.U., Landsberger, N., Strouboulis, J. &
     Wolffe, A.P. (1998) Methylated DNA and MeCP2 recruit histone deacetylase to repress tran-
     scription. *Nat. Genet.* **19**, 187–191

26. Ng, H.H., Zhang, Y., Hendrich, B., Johnson, C.A., Turner, B.M., Erdjument-Bromage, H., Tempst,
     P., Reinberg, D. & Bird, A. (1999) MBD2 is a transcriptional repressor belonging to the MeCP1
     histone deacetylase complex. *Nat. Genet.* **23**, 58–61

27. Nan, X., Campoy, F.J. & Bird, A. (1997) MeCP2 is a transcriptional repressor with abundant bind-
     ing sites in genomic chromatin. *Cell* **88**, 471–481

28. Hendrich, B., Hardeland, U., Ng, H.H., Jiricny, J. & Bird, A. (1999) The thymine glycosylase MBD4
     can bind to the product of deamination at methylated CpG sites. *Nature (London)* **401**, 301–304

29. Bhattacharya, S.K., Ramchandani, S., Cervoni, N. & Szyf, M. (1999) A mammalian protein with specific demethylase activity for mCpG DNA. *Nature (London)* **397**, 579–583

30. Chandler, S.P., Guschin, D., Landsberger, N. and Wolffe, A.P. (1999) The methyl-CpG binding transcriptional repressor MeCP2 stably associates with nucleosomal DNA. *Biochemistry* **38**, 7008–7018

31. Bird, A.P. & Wolffe, A.P. (1999) Methylation-induced repression — belts, braces, and chromatin. *Cell* **99**, 451–454

32. Zhang, Y., Ng, H.H., Erdjument-Bromage, H., Tempst, P., Bird, A. & Reinberg, D. (1999) Analysis of the NuRD subunits reveals a histone deacetylase core complex and a connection with DNA methylation. *Genes Dev.* **13**, 1924–1935

33. Okano, M., Xie, S. & Li, E. (1998) Cloning and characterization of a family of novel mammalian DNA (cytosine-5) methyltransferases. *Nat. Genet.* **19**, 219–220

34. Okano, M., Bell, D.W., Haber, D.A. & Li, E. (1999) DNA methyltransferases Dnmt3a and Dnmt3b are essential for *de novo* methylation and mammalian development. *Cell* **99**, 247–257

35. Razin, A. & Kafri, T. (1994) DNA methylation from embryo to adult. *Prog. Nucleic Acid Res. Mol. Biol.* **48**, 53–81

36. Mayer, W., Niveleau, A., Walter, J., Fundele, R. & Haaf, T. (2000) Demethylation of the zygotic paternal genome. *Nature (London)* **403**, 501–502

37. Mertineit, C., Yoder, J.A., Taketo, T., Laird, D.W., Trasler, J.M. & Bestor, T.H. (1998) Sex-specific exons control DNA methyltransferase in mammalian germ cells. *Development* **125**, 889–897

38. Macleod, D., Clark, V. & Bird, A.P. (1999) Absence of genome-wide changes in DNA methylation during development of the zebrafish (Danio rerio). *Nat. Genet.* **23**, 139–140

39. Kinoshita, K., Bessho, T. & Asashima, M. (1993) Competence prepattern in the animal hemisphere of the 8-cell-stage Xenopus embryo. *Dev. Biol.* **160**, 276–284

40. Wolf, E., Zakhartchenko, V. & Brem, G. (1998) Nuclear transfer in mammals: recent developments and future perspectives. *J. Biotechnol.* **65**, 99–110

41. Gurdon, J.B., Laskey, R.A. & Reeves, O.R. (1975) The developmental capacity of nuclei transplanted from keratinized skin cells of adult frogs. *J. Embryol. Exp. Morphol.* **34**, 93–112

6

Signalling from the cell surface to the nucleus

Melanie Lee* and Stephen Goodbourn†[1]

*Marie Curie Cancer Research Institute, The Chart, Oxted, Surrey, RH8 OTL, U.K., and †Department of Biochemistry and Immunology, St. George's Hospital Medical School, University of London, London SW17 0RE, U.K.

Introduction

Changes in the patterns of specific gene expression in response to external stimuli can lead to multiple outcomes, such as growth, inhibition of growth, differentiation, immune protection or apoptosis, and enable a multicellular organism to co-ordinate its response to complex environmental signals. These changes are often executed through the regulation of transcriptional initiation, brought about by alterations in the properties of promoter-specific transcription factors which ultimately affect the efficiency of recruitment of RNA polymerase to the target gene. The mechanisms of these latter events are the subject of other chapters in this volume. Here we review our knowledge of how the perception of a ligand binding to a cell-surface receptor is transduced into a signal that affects the promoters of an often limited number of genes.

Ligands and receptors

Extracellular ligands exert their effects on target cells by binding with high affinity to receptors. Most types of receptor span the membrane and detect the ligand outside the cell. The exceptions to this are steroids, retinoids, vitamin D

[1]To whom correspondence should be addressed
(e-mail: s.goodbourn@sghms.ac.uk).

and nitric oxide, which cross the cytoplasmic membrane and interact with intracellular receptors.

There are three main classes of cell-surface receptor. The first to be discovered were the ion channels. However, since these play a relatively minor role in regulating the patterns of gene expression they will not be considered further. The second are the G-protein-coupled receptors and the third are the enzyme-linked receptors. By far the best studied of the latter class are the receptor tyrosine kinases (e.g. the receptors for polypeptide growth factors such as insulin, epidermal growth factor and platelet-derived growth factor [PDGF]), which, as their name implies, have the ability to phosphorylate substrate molecules on specific tyrosines. However, there are several other types of receptor that are linked to enzymic pathways. Some recruit non-receptor tyrosine kinases (e.g. the receptors for interferons and many interleukins), whereas others recruit tyrosine phosphatases (e.g. the erythropoietin receptor). In at least one case (transforming growth factor β and related receptors), ligand binding activates receptor serine/threonine kinase function. Finally, there are some receptors that may be linked to specific proteolysis events (e.g. the Notch receptor).

G-protein-coupled receptors

Receptors that are linked to G-proteins [1] all have the common structural feature of being a single polypeptide that spans the cytoplasmic membrane seven times (the 'seven-pass receptors') (Figure 1a). The ligand-binding domain is on the outside of the cell whereas the effector domain is on the cytoplasmic side of the membrane. Upon ligand binding, a conformational change occurs such that the effector domain interacts with a specific membrane-bound heterotrimeric G-protein. Prior to receptor binding, the Gα subunit has a tightly bound GDP molecule. The association with the activated receptor causes the affinity of the Gα subunit for GDP to decrease and its affinity for GTP to increase. The Gα–GTP complex loses its ability to interact with the Gβγ subunits and is released from the receptor complex. The separated G-protein components can initiate a variety of downstream signalling events.

The G-protein-coupled receptors are classified according to the type of Gα protein they activate (e.g. $G_s\alpha$, $G_i\alpha$, $G_q\alpha$, and $G\alpha_{12}$). Each subtype of Gα–GTP can execute a distinct type of downstream signaling and the details of two important classes of Gα protein are summarized in Figure 1(b). For $G_s\alpha$-linked receptors (e.g. the receptors for adrenaline and vasopressin), the Gα–GTP binds to and activates membrane-bound adenylate cyclase, which in turn produces cAMP from ATP [1]. For $G_i\alpha$-linked receptors (e.g. α_2-adrenergic receptors), the Gα–GTP binds to and inhibits membrane-bound adenylate cyclase [1]. cAMP binds to the dimeric regulatory subunit of protein kinase A (PKA), causing it to dissociate from the dimeric catalytic subunit. Activated PKA enters the nucleus and phosphorylates a sequence-specific

(a)

(b)

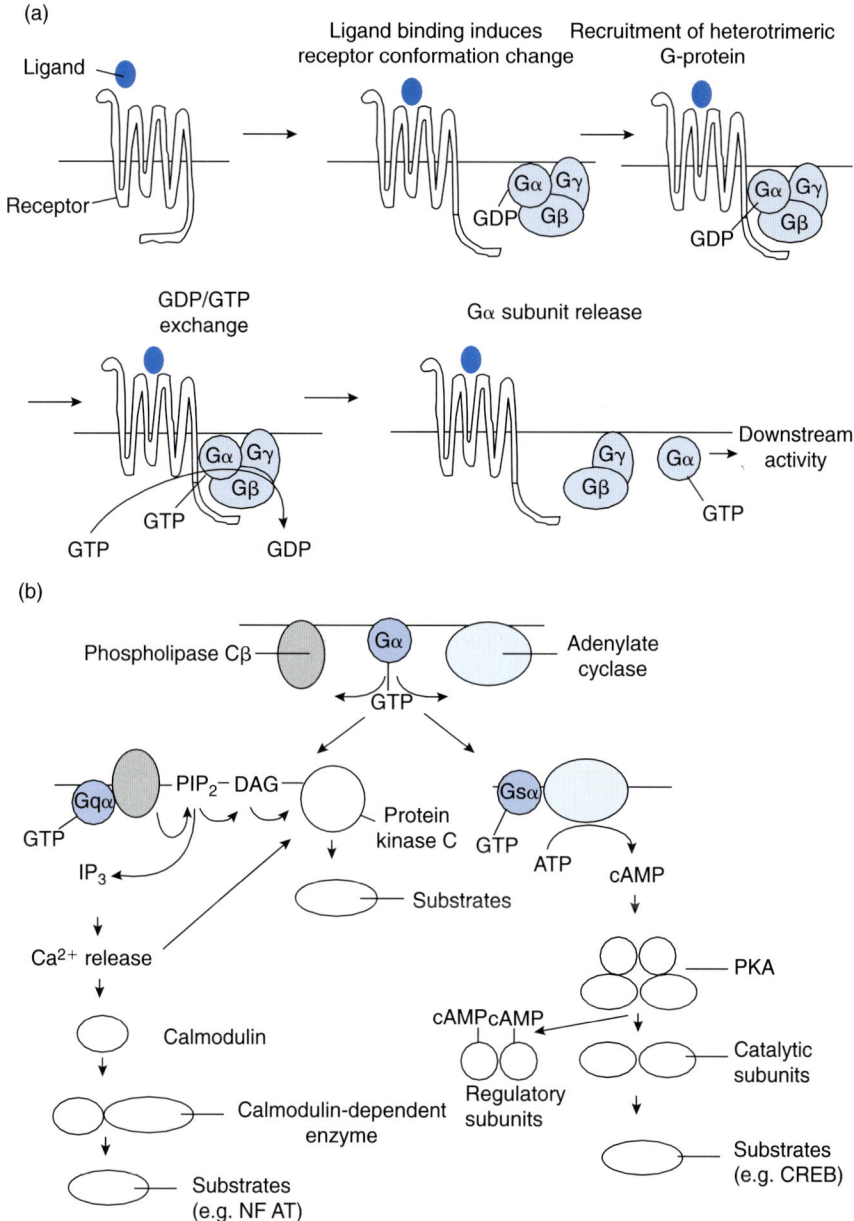

Figure 1. G-protein-coupled receptors

(a) Ligand binding induces a conformational change in the effector domain of the G-protein-coupled receptor, allowing recruitment of a specific heterotrimeric G-protein. The Gα subunit of the G-protein consequently becomes GTP-bound and dissociates from the Gβγ subunit, initiating downstream events. (b) The left-hand side of the diagram shows that $G_q\alpha$–GTP binds to and activates phospholipase Cβ, which catalyses the conversion of PIP_2 into IP_3 and DAG. These effectors activate downstream signalling pathways as indicated, and lead to the activation of transcription factors as discussed in the text. The right-hand side of the diagram shows that $G_s\alpha$–GTP binds to and activates adenylate cyclase, thereby stimulating cAMP production. cAMP binds to and dissociates from the regulatory subunits of PKA, allowing the catalytic subunits to modify transcription factors such as CREB.

factor [called cAMP-response-element-binding protein (CREB)] that is pre-bound to the promoters of target genes. CREB is then able to interact with another protein called CBP (CREB-binding protein) [2]. In turn, CBP makes multiple contacts with proteins in the basal transcription machinery and also helps to 'remodel' the chromatin to enable RNA polymerase to be recruited.

For the $G_q\alpha$ class (e.g. α-adrenergic receptors), $G\alpha$–GTP binds to and activates membrane-bound phospholipase Cβ (Figure 1b) [1]. This catalyses the hydrolysis of a rare membrane phospholipid, phosphatidylinositol 4,5-bis-phosphate (PIP$_2$), giving rise to diacylglycerol (DAG), which remains mem-brane-bound, and myo-inositol 1,4,5-trisphosphate (IP$_3$) which is soluble. The IP$_3$ binds to specific receptors on the endoplasmic reticulum, causing release of stored Ca^{2+}. In turn, Ca^{2+} binds to and activates calmodulin, an essential cofactor for a variety of specific protein kinases and phosphatases, e.g. the transcription factor NFAT (nuclear factor of activated T-cells), which enters the nucleus as a result of dephosphorylation by a calmodulin-dependent phos-phatase [3]. DAG can be further metabolized to signalling molecules such as arachidonic acid, or it can bind to and activate membrane-bound protein kinase C, an enzyme that plays pleiotropic roles in gene activation [4].

It is becoming clear that $G\alpha$ subunits can also signal to MAPK (mitogen-activated protein kinase) modules (discussed in more detail later) in a number of novel ways [5]. The mechanisms of this signalling are only just beginning to be understood.

Receptor tyrosine kinases

The specific binding of ligand induces dimerization and juxtapositioning of receptor subunits. Either as a result of this or as a result of dimerization-induced conformational changes, the innate tyrosine kinase potential of the receptor is activated, leading to the reciprocal phosphorylation of receptor chains [6]. This scheme of activation is outlined in Figure 2(a). In cases where the receptor has no innate tyrosine kinase activity, the ligand-induced dimerization leads to the activation of receptor-associated tyrosine kinases, which phosphorylate the receptor subunits. For example, when type II interferon binds to its heterodimeric receptor, the tyrosine kinase [Janus kinase 1 (JAK1)] associated with the receptor β chain, becomes active and phosphorylates the receptor α chain and its associated tyrosine kinase (JAK2). JAK2 in turn phosphorylates the receptor β chain and further activates JAK1 [7].

The phosphorylated tyrosine residues are the target for proteins that con-tain an Src homology 2 (SH2) domain, so called because of a homology to a region of c-Src, a prototypical receptor-associated tyrosine kinase [8]. Although SH2 domains recognize phosphotyrosine, there is a considerable degree of specificity in this. There are several types of SH2 domain protein (many containing multiple SH2 domains). Since activated receptors contain

(a)

(b)

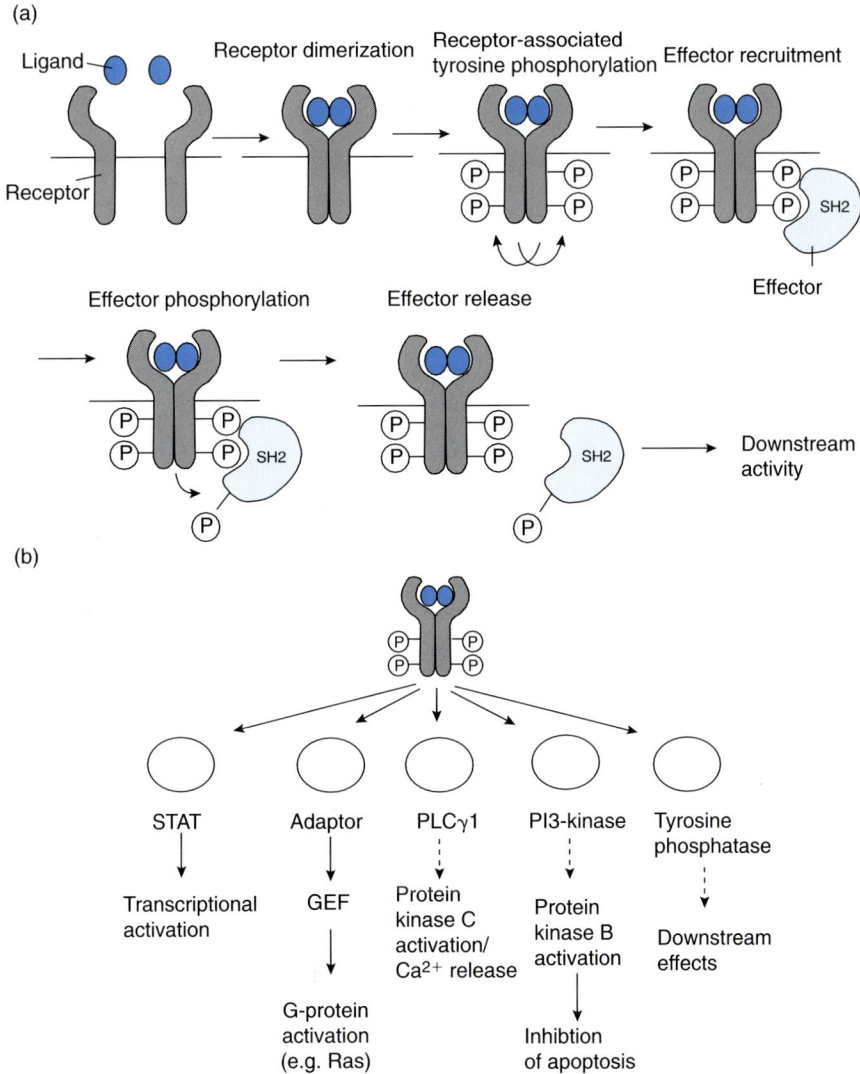

Figure 2. Receptor tyrosine kinases
(a) Binding of ligand induces receptor dimerization and consequent reciprocal phosphorylation of specific tyrosine residues in the receptor chains. The phosphorylated tyrosines are recognized by SH2 (Src homology 2)-domain proteins (effectors), which are recruited to the activated receptor, become phosphorylated and initiate downstream signalling. (b) Classes of SH2-domain factors which can be recruited to activated receptors are illustrated, together with downstream effector molecules. PI 3-kinase, phosphoinositide 3-kinase; PLCγ, phospholipase Cγ; Solid arrows, direct effects; broken arrows, indirect effects.

many phosphotyrosines, they appear to have the potential to activate several pathways, raising the unresolved issues of how specificity is achieved. The classes of SH2-domain factors that have been shown to interact with activated receptors are summarized in Figure 2(b).

The most direct SH2-linked pathway that can lead to specific gene expression involves the STATs (signal transduction and activators of transcription) [7]. Initially, two STAT molecules bind independently as monomers to the phosphorylated receptors through their SH2 domains and are thus brought into proximity with the tyrosine kinases, which in turn phosphorylate them. This causes a conformational change in the STATs such that their SH2 domains now bind to the second phosphorylated STAT, creating a dimer. The dimerized STATs dissociate from the receptor and migrate to the nucleus, where they can bind to DNA directly in a sequence-specific manner or associate with heterologous DNA-binding proteins to bind to other promoter sequences.

Another major class of SH2 proteins that interact with activated receptors are the adaptor proteins, the best characterized of which is growth-factor-receptor-bound protein Grb2, has no associated kinase activity. Grb2 contains motifs that mediate interactions with other factors, for example SOS (son of sevenless), thus serving as a molecular recruitment agent [9]. SOS is a member of a family of factors called GDP/GTP exchange factors (GEFs), which, once activated, can catalyse the exchange of GDP for GTP on monomeric G-proteins (see later).

Phospholipase Cγ1 also contains an SH2 domain and, like its counterpart PLCβ discussed above, hydrolyses PIP_2 to DAG and IP_3, leading to a rise in Ca^{2+} levels and protein kinase C activation. Receptor stimulation can also lead to the recruitment and activation of phosphoinositide 3-kinase which phosphorylates membrane PIP_2 generating phosphatidylinositol 3,4,5-trisphosphate (PIP_3). PIP_3 then recruits target proteins to the membrane and activates them by juxtaposing them to activating kinases. The best characterized target protein is protein kinase B [10], which plays a major role in protecting cells from apoptosis. Finally, protein tyrosine phosphatases also contain SH2 domains, and can bind to activated receptors and presumably down-regulate gene expression.

It is clear from the above that receptor tyrosine kinases can potentially initiate many downstream signalling events. The activation of any given receptor can often elicit multiple independent interactions e.g. PDGF receptor becomes phosphorylated on at least six tyrosines, each of which interacts with distinct SH2-domain target proteins [11], each of which could lead to distinct gene activations. However, this might not necessarily be so. For example, binding of PDGF to its receptor can activate phosphoinositide 3-kinase either by direct binding or via the intermediacy of Ras–GTP. To evaluate the relative contributions of the different pathways initiated from the PDGF receptor, Famborough et al. [12] created receptor variants containing distinct tyrosine to phenylalanine changes and introduced these into NIH3T3 mouse fibroblasts. The cells were then stimulated with PDGF and the gene expression profiles analysed. Taking advantage of the enormously powerful chip hybridization technologies they were able to determine a gene expression profile of the different cell lines. Surprisingly, when individual signalling pathways were 'knocked out', only a very small number of genes failed to be expressed in response to PDGF. Rather, there were quantitative differences in expression for any given target, suggesting that gene activation is mediated by a combination of weak signals that builds specificity.

The MAPK pathways

Perhaps the most important effectors recruited by receptor tyrosine kinases are the GEFs, which recognize a specific monomeric G-protein, promote the exchange of GDP for GTP and change the interaction spectrum of the G-protein. The best characterized of the target monomeric G-proteins is the Ras protein [13], which is a member of a large family. Figure 3 shows that the GTP-bound form of Ras has multiple downstream targets, including a protein called Raf. Ras is always membrane bound, and thus when it becomes modified by GTP association it specifically recruits Raf to the membrane, where Raf is activated in a manner not yet understood. Activated Raf is a protein kinase, its target being another protein kinase, MEK-1, which is the kinase MAPK/ERK (ERK stands for extracellular signal-related kinase). MEK-1 in turn phosphorylates and activates ERK, which can phosphorylate transcription factors leading to the alteration of transcription. For example, in response to serum treatment after starvation, fibroblast cells activate ERK which phosphorylates Elk-1. Elk-1 is then able to bind to the promoter of the c-*fos* gene in a co-operative manner with the DNA-binding protein SRF (serum response factor) [14].

Kinase cascades are a common theme in signal transduction [15] and the pathway leading from Ras activation to Elk-1 phosphorylation and c-Fos transcriptional induction is an example of a 'MAPK cascade'. These are widely used in biological systems and serve as a good example of how signal specificity can be achieved (Figure 3). ERK is a MAPK, of which there are at least 12 in mammals, whilst MEK-1 is a MAPK kinase (MAPKK), of which there are at least seven in mammals. The MAPKKs are dual-specificity kinases, phosphorylating both a threonine residue and a tyrosine residue that are closely linked in their MAPK substrates. The MAPKKs show marked substrate specificity. However, the enzymes that activate the MAPKKs, the MAPKK kinases (MAPKKKs), such as Raf, are diverse in structure, are more numerous than either the MAPKs or MAPKKs (there are at least fourteen in mammals) and, unlike the MAPKKs, often have a broad specificity. The diversity in structure in the MAPKKKs usually resides outside their catalytic and downstream signalling functions, and these diversified regions often contain multiple motifs that permit interaction with other proteins.

Taken together, the above observations raise the potential problem of a breakdown in specificity, namely that stimulation through different receptors could potentially activate the same MAPKKs through distinct MAPKKKs

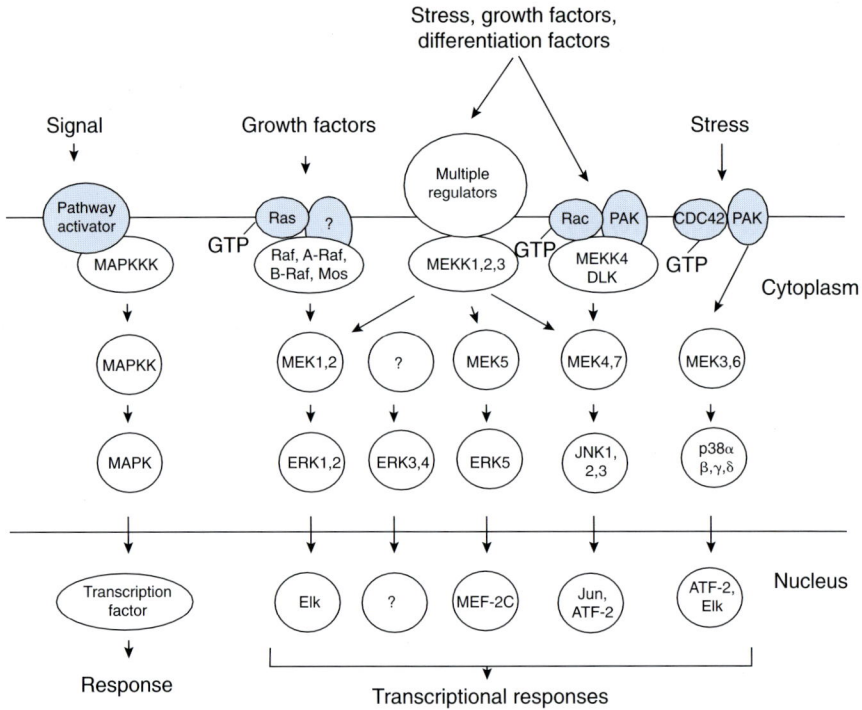

Figure 3. MAPK signalling pathways in mammalian cells
A summary of the currently identified interactions between components of the mammalian MAPK cascade pathways is illustrated. On the left is the generalized signalling pathway. The specific components involved in the pathways activated by specific signals are shown on the right. ATF, activating transcription factor; CDC, cell division cycle; DLK, dual lineage kinase; JNK, c-Jun N-terminal kinase; MEKK, MEK kinase; MEF, myocyte enhancer factor; PAK, p21-activated protein kinase.

(Figure 3). In fact, receptor cross-talk is avoided and specificity is tightly maintained even though some individual kinases appear to be used in more than one signal transduction pathway. The best understanding of the mechanisms by which this is mediated comes from work on *Saccharomyces cerevisiae*, which can exist in a vegetative haploid or diploid state or undergo filamentation depending on its environment. The decision over which form the yeast will take depends on signal transduction pathways which involve several common components.

Haploid yeast are one of two mating types, α or **a**, each of which secretes a simple polypeptide pheromone that is received by a receptor on the other mating type. The receptors are members of the seven-pass family, and upon ligand binding recruit a heterotrimeric G-protein and catalyse the exchange of GDP for GTP on the Gα subunit. Unlike the examples discussed above, it is the Gβγ subunit complex (called Ste4p and Ste18p) that carries out the downstream signalling. The liberated Ste4p–Ste18p complex remains membrane bound and is now free to interact with Ste5p, which is recruited to the membrane as a result of this interaction. Ste5p is a 'scaffold protein' that contains binding sites for Ste11p, Ste7p and Fus3p, which are respectively the MAP-KKK, the MAPKK and the MAPK known to be essential for the yeast mating response. Although the entire MAPK module is thus recruited as a supercomplex to the pathway-specific initiator Ste4p, a phosphorylation event is needed to activate the module. A crucial component for this is Ste20p, a membrane-bound kinase that can also interact with Ste4p. In combination with Ste50p, Ste20p phosphorylates and activates the MAPKKK (Ste11p), which is in its proximity by virtue of being assembled on the Ste5p scaffold (Figure 4a). Once Ste11p is activated, the other enzymes in the cascade become activated. The ultimate phosphorylation of Fus3p causes it to lose the ability to interact with the MAPKK (Ste7p), and it enters the nucleus to find transcription factor targets [16]. Thus Ste5p can be considered to impose the specificity on the pathway since it links the pathway-specific initiator, Ste4p, to the pathway-specific MAPK, Fus3p.

The importance of the scaffold is indicated by the fact that the filamentation response of yeast also uses Ste20p, Ste50p, Ste11p and Ste7p, but targets the signal to a distinct MAPK, Kss1p (Figure 4a). A scaffold protein unique to this pathway has not yet been detected, but it is not Ste5p. Ste11p is also used in a pathway that turns on the transcription of genes required for glycerol biosynthesis in response to osmotic shock (Figure 4a). In this pathway, Ste11p is recruited into a complex with a different MAPKK (Pbs2p) and MAPK (Hog1p). Pbs2p seems to serve as the pathway-specific scaffold, since it interacts with both upstream and downstream kinases as well as with the pathway-specific initiator Sho1p (an osmosensor). Once again the activating phosphorylation event comes from Ste20p and Ste50p. Since Pbs2p interacts with Sho1p rather than Ste4p/Ste18p and, with Hog1p rather than Fus3p it specifically links input to output despite using some components in common with the pheromone response pathway.

MAPK module scaffold proteins have recently been detected in mammalian cells (Figure 4b) [15]. For example, the protein JIP-1 interacts with a mammalian Ste20p homologue (HPK-1), a MAPKKK (MLK3), a MAPKK (MEK7), and a MAPK [JNK (c-Jun N-terminal kinase)] and thus like Ste5p may link a specific input signal to the activation of transcription factors such as c-Jun. JNK may also be activated by distinct signals in a different module containing JNK, the MAPKK, MEK4, and the MAPKKK protein, MEKK1 (MEK kinase 1), which appears to act as a scaffold as well as an enzyme in this pathway. MEKK1 can also bind to potential input signals, such as the Ste20p homologue, NIK (Nck-interacting kinase). Similar to the activation of JNK, scaffold molecules may exist to specify ERK activation. The MP1 (MEK partner 1) protein appears to link ERK1 to MEK1 and possibly to activation by Raf and thus Ras–GTP, whereas KSR1 (kinase suppressor of Ras1) links Ras–GTP to MEK1 and ERK2. As a result of these types of molecules the cross-talk anticipated in Figure 3 may be avoided.

   MAPKs show a remarkable degree of substrate specificity, in many cases
binding directly to their target transcription factor. For example, Kss1p can
bind to Ste12p, JNK can bind to c-Jun and activating transcription factor 2
(ATF-2), and ERK can bind to Elk-1 (Figure 4). However, this is clearly insufficient to account for all of the pathway specificity since the distinct patterns of
transcription associated with the yeast mating and filamentation responses are
both mediated by MAPK-activated Ste12p; however, Ste12p activated by
Fus3p cannot stimulate transcription of the filamentation response genes, and
Ste12p activated by Kss1p cannot normally stimulate transcription of the
pheromone response genes [17]. The probable answer to this is that Ste12p is
bound to the promoters of the two classes of target genes as a component of

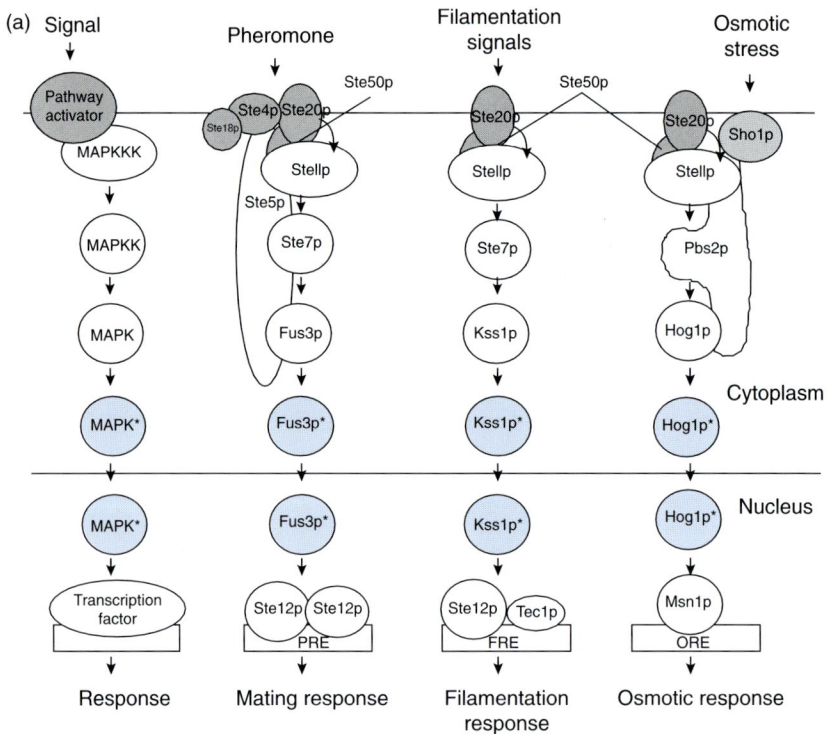

Figure 4. MAPK signalling in S. cerevisiae
(a) In S. cerevisiae the transcriptional response to mating pheromone, filamentation signals and
osmotic stress are mediated by common and distinct components of MAPK signalling pathways.
On the left is the generalized signalling pathway. The specific components involved in the pathways activated by specific signals are shown on the right. Specificity is achieved through 'scaffold'
proteins such as Ste5p and Pbs2p, which link the pathway-specific initiator to the pathway-specific MAPK as illustrated. The activated MAPK in each pathway is shown in blue. This molecule
passes into the nucleus to activate transcription. FRE, filamentation response element; ORE,
osmoregulatory response element; PRE, pheromone response element. The asterisks indicate
that the relevant protein kinase has been activated as a consequence of the upstream event.

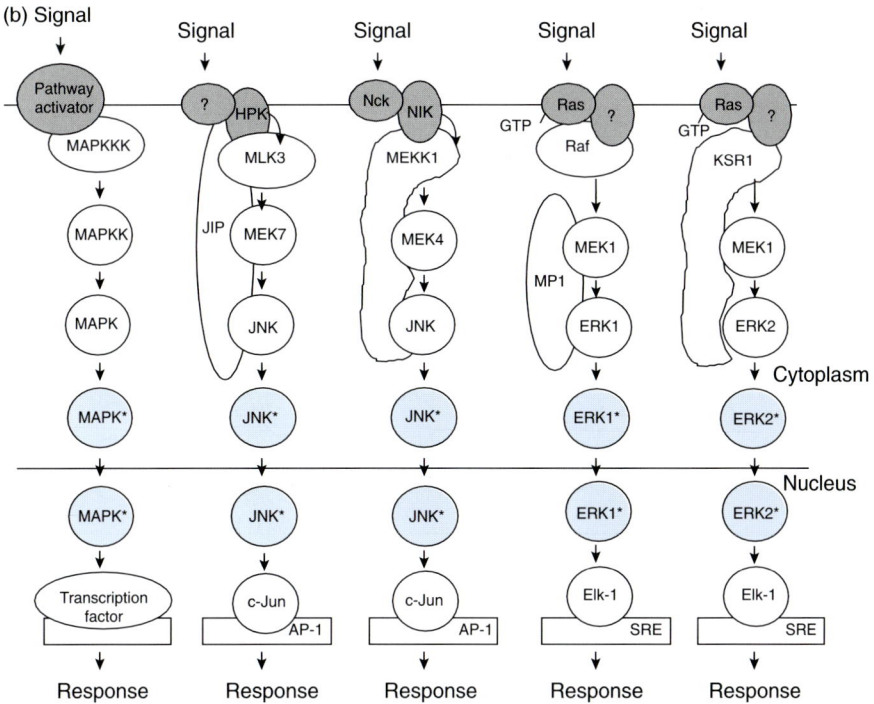

(b) Putative scaffolded complexes may insulate distinct MAPK pathways in mammalian cells. Recently identified putative MAPK modules are illustrated. On the left is the generalized signalling pathway. The specific components involved in the pathways activated by individual signals are shown on the right. The activated MAPK in each pathway is shown in blue. This molecule passes into the nucleus to activate transcription. AP-1, activator protein-1; HPK, haematopoietic progenitor kinase; JIP, JNK interacting protein-1; KSR, kinase suppressor of Ras; MLK, mixed lineage kinase; MPI, MEK partner 1; NIK, Nck-interacting kinase.; SRE, serum response element The asterisks indicate that the relevant protein kinase has been activated as a consequence of the upstream event.

distinct protein complexes and may therefore be differentially accessible to MAPKs: Ste12p binds as a homodimer to a pheromone response element and as a heterodimer with Tec-1p to a filamentation response element.

Proteolysis

Although most pathways involve alterations in the phosphorylation state of components, other covalent modifications can occur in which transcription factors or their regulatory partners are proteolytically processed [18]. The transcription factor nuclear factor κB (NF-κB) is required for the activation of a number of immediate immune response genes, such as those encoding cytokines and attachment molecules [19]. Prior to stimulation, NF-κB is retained in the cytoplasm by association with an inhibitor molecule, IκB (Figure 5). Treatment of cells with a range of external stimuli, e.g. interleukin-1,

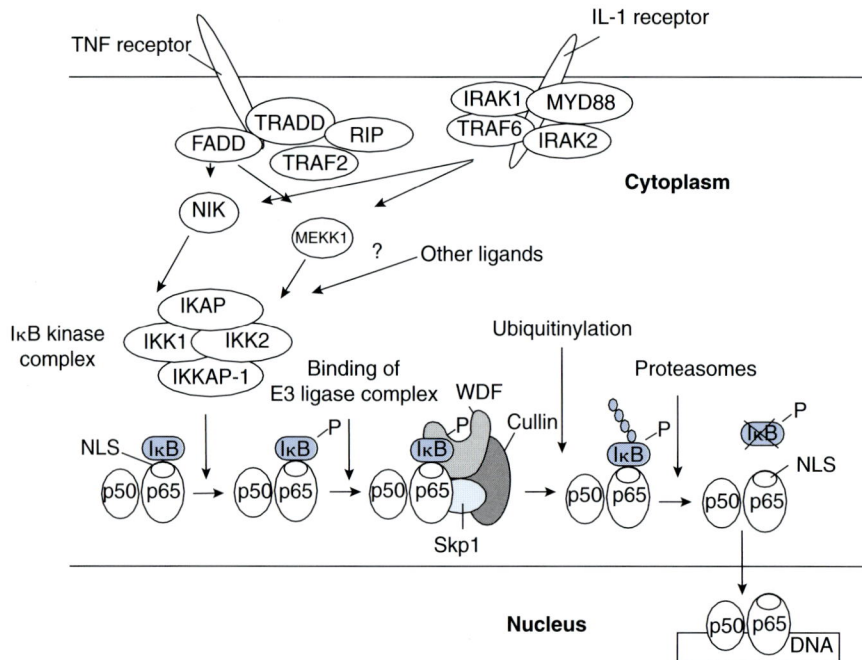

Figure 5. Activation of NF-κB is mediated by proteolytic degradation of IκB
Ligand binding to the tumour necrosis factor (TNF) or interleukin (IL)-1 receptors induces signalling pathways that feed into a multi-component IκB kinase complex, which, once activated,
phosphorylates IκB. The phosphorylated IκB is recognized and bound by a target-specific E3
ubiquitin ligase complex, resulting in the ubiquitinylation and proteolytic destruction of IκB. The
liberated NF-κB enters the nucleus and regulates the transcription of target genes. FADD, Fas-
associated death domain protein; IKAP, IKK complex-associated protein; IKK, IκB kinase; IKKAP,
IKK-associated protein; IRAK, interleukin-1 receptor-associated kinase; MYD, myeloleukaemic
differentiation protein; NIK, NF-κB-inducing kinase (N.B. this is a distinct protein from the Nck-
interacting kinase discussed in Figure 4); NLS, nuclear localization signal; RIP, TNF receptor-
interacting protein; TRADD, TNF receptor-associated death domain protein; TRAF, TNF receptor-associated factor; WDF, WD40 repeat/F box protein.

tumour necrosis factor, bacterial lipopolysaccharides, leads to a dramatic
increase in the nuclear levels and effective DNA-binding concentration of NF-
κB. The biochemical events which precede this are complex, but involve a
specific phosphorylation of IκB. One of the more puzzling aspects of NF-κB
activation has concerned the ability of so many different types of ligand to
cause activation, but it now appears that this is brought about by having the
disparate signals feed into a multi-component kinase which contains its own
scaffold molecules. For example, TNF receptor occupation recruits and
activates a MAPKKK called NIK (NF-κB-inducing kinase; this is not the same
protein as the MEKK1 activator discussed above), that in turn activates IKK1,
a kinase in the IκB kinase complex. The scaffold protein, IKKAP-1 (IKK-
associated protein-1), may directly contact the TNF receptor to facilitate
activation.

Phosphorylated IκB is recognized by a receptor (an 'F box/WD40 repeat' protein, belonging to the recently identified β-TrCP/Slimb family) which is a component of a target-specific E3 ubiquitin ligase [20]. The combined complex is then recognized by adaptor proteins (of the Skp1 and cullin groups) and, in conjunction with an E1 ubiquitin-conjugating enzyme and an E2 ubiquitin ligase, the phosphorylated IκB becomes ubiquitinylated on two lysine residues. The ubiquitinylated IκB protein is thus targeted to multi-catalytic proteasomes which ensure its destruction. Freed from inhibitor, the nuclear localization signal of NF-κB is unmasked and the protein can enter the nucleus unencumbered.

Regulation of transcription factor function by proteolysis is not restricted to NF-κB/IκB. The Wingless pathway plays a crucial role in development in *Drosophila melanogaster*, and the Wnt pathway may play a similar role in mammals (Figure 6) [21]. The ultimate consequence of activation of these path-

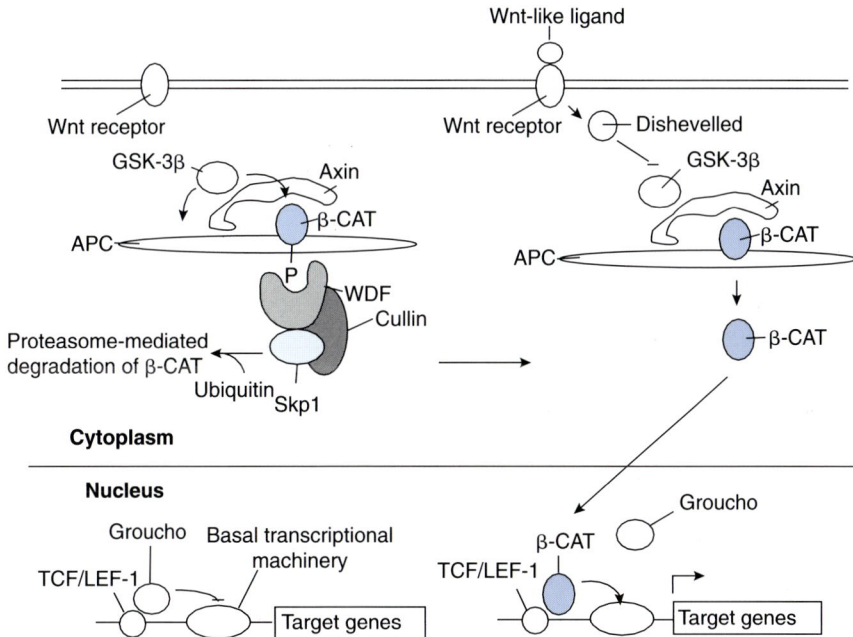

Figure 6. Activation of the Wingless/Wnt pathway leads to stabilization of β-catenin and transcriptional activation
Prior to stimulation, β-catenin forms a multi-protein complex with Axin and the APC gene product and is phosphorylated by GSK-3β. This phosphorylation targets β-catenin for proteasome-mediated degradation via ubiquitination using a specific E3 ubiquitin ligase complex. Upon receipt of a Wnt-like ligand, GSK-3β is inactivated using an ill-defined mechanism involving the product of the *dishevelled* gene. The β-catenin becomes dephosphorylated by cellular phosphatases and is released, whereupon it can enter the nucleus and is recruited to promoters via an interaction with the DNA-binding protein TCF/LEF-1. This interaction displaces transcriptional co-repressors such as the product of the Groucho gene. APC, adenomatous polyposis coli; β-CAT, β-catenin; WDF, WD40 repeat/F box protein.

ways is the modification of the transcriptional output of target cells. Stimulation of the Wnt pathway leads to an increase in β-catenin levels, without an apparent need for *de novo* protein synthesis, suggesting that signalling alters the stability of β-catenin [22,23]. In fact, β-catenin is normally turned over by the same process that inactivates IκB, namely site-specific phosphorylation, followed by site-specific ubiquitinylation utilizing an F box–Skp1–Cullin complex, followed ultimately by proteasome-mediated degradation. In contrast to IκB, β-catenin is constitutively phosphorylated and therefore constitutively turned over. The kinase responsible is glycogen synthase kinase (GSK3β) which is specifically targeted to β-catenin by a scaffold protein (Axin) that interacts with a complex containing both β-catenin and the product of the proto-oncogenic adenomatous polyposis coli (APC) gene. Upon receipt of the Wnt/wingless signal, GSK-3β is inhibited by the product of the dishevelled gene, which also gets recruited to the Axin–APC–β-catenin complex. Once the constitutive phosphorylation ceases, cellular phosphatases rapidly dephosphorylate β-catenin, which can dissociate from the complex and is no longer degraded. β-catenin accumulates, passes into the nucleus and interacts with DNA-bound LEF/TCF and displaces a transcriptional co-repressor (Groucho) normally associated with this factor to de-repress target genes [24].

Summary

- *Transcriptional initiation is regulated by altering the properties of promoter-specific DNA-binding proteins, such that these proteins either show altered interaction with the basal transcriptional machinery, or show changes in their cytoplasmic/nuclear distribution.*
- *Information is passed from the receptor to the transcription factor by a process of post-translational modifications of pathway components.*
- *Post-translational modifications can include phosphorylations or dephosphorylations, which are reversible, and proteolysis, which is irreversible.*
- *Specificity within a linear signal transduction pathway is preserved by the existence of scaffold proteins, which serve to co-localize many of the factors from a given linear pathway.*

References

1. Gutkind, J.S. (1998) Cell growth control by G protein-coupled receptors: from signal transduction to signal integration. *Oncogene* **17**, 1331–1342
2. Montminy, M. (1997) Transcriptional regulation by cyclic AMP. *Annu. Rev. Biochem.* **66**, 807–822
3. Crabtree, G.R. (1999) Generic signals and specific outcomes: signaling through Ca^{2+}, calcineurin, and NF-AT. *Cell* **96**, 611–614
4. Mellor, H. & Parker, P.J. (1998) The extended protein kinase C superfamily. *Biochem. J.* **332**, 281–292

5.   Luttrell, L.M., Daaka, Y. & Lefkowitz, R.J. (1999) Regulation of tyrosine kinase cascades by G-protein-coupled receptors. *Curr. Opin. Cell. Biol.* **11**, 177–183

6.   Porter, A.C. & Vaillancourt, R.R. (1998) Tyrosine kinase receptor-activated signal transduction pathways which lead to oncogenesis. *Oncogene* **17**, 1343–1352

7.   Stark, G.R., Kerr, I.M., Williams, B.R., Silverman R.H. & Schreiber, R.D. (1998) How cells respond to interferons. *Annu. Rev. Biochem.* **67**, 227–264

8.   Sawyer, T.K. (1998) Src homology-2 domains: structure, mechanisms, and drug discovery. *Biopolymers* **47**, 243–261

9.   Buday, L. (1999) Membrane-targeting of signaling molecules by SH2/SH3 domain-containing adaptor proteins. *Biochim. Biophys. Acta* **1422**, 187–204

10. Downward, J. (1998) Mechanisms and consequences of activation of protein kinase B/Akt. *Curr. Opin. Cell. Biol.* **10**, 262–267

11. Pawson, T. & Saxton, T.M. (1999) Signaling networks - do all roads lead to the same genes? *Cell* **97**, 675–678

12. Famborough, D., McClure, K., Kazlauskas, A. & Lander, E.S. (1999) Diverse signaling pathways activated by growth factor receptors induce broadly overlapping rather than independent sets of genes. *Cell* **97**, 727–741

13. Wittinghofer, A. (1998) Signal transduction via Ras. *Biol. Chem.* **379**, 933–937

14. Treisman, R., Alberts, A.S. & Sahai, E. (1998) Regulation of SRF activity by Rho family GTPases. *Cold Spring Harbor Symp. Quant. Biol.* **63**, 643–651

15. Garrington, T. & Johnson, G.L. (1999) Organisation and regulation of mitogen-activated protein kinase signaling pathways. *Curr. Opin. Cell Biol.* **11**, 211–218

16. Reiser, V., Ammerer, G. & Ruis, H. (1999) Nucleocytoplasmic traffic of MAP kinases. *Gene Expression* **7**, 247–254

17. Madhani, H.D., Styles, C.A. & Fink, G.R. (1997) MAP kinases with distinct inhibitory functions impart signaling specificity during yeast differentiation. *Cell* **91**, 673–684

18. Goodbourn, S. & King, P. (1997) Regulation of transcription factor function by proteolysis. *Biochem. Soc. Trans.* **25**, 498–502

19. Mercurio, F. & Manning, A.M. (1999) Multiple signals converging on NF-κB. *Curr. Opin. Cell Biol.* **11**, 226–232

20. Maniatis, T. (1999) A ubiquitin ligase complex essential for the NF-kappaB, Wnt/Wingless, and Hedgehog signaling pathways. *Genes Dev.* **13**, 505–510

21. Dale, T. (1998) Signal transduction by the Wnt family of ligands. *Biochem. J.* **329**, 209–223

22. Polakis, P. (1999) The oncogenic activation of β-catenin. *Curr. Opin. Cell Biol.* **9**, 15–21

23. Eastman, Q. & Grosschedl, R. (1999) Regulation of LEF-1/TCF transcription factors by Wnt and other signals. *Curr. Opin. Cell Biol.* **11**, 233–240

24. Nusse, R. (1999) WNT targets. Repression and activation. *Trends Genet.* **15**, 1–3

7

Control of gene expression and the cell cycle

Ho Man Chan, Noriko Shikama and Nicholas B. La Thangue[1]

Division of Biochemistry and Molecular Biology, Davidson Building, University of Glasgow, Glasgow G12 8QQ, U.K.

Introduction

During the mammalian cell cycle the transcription of a large group of genes is integrated with the G_1 to S phase transition, thus providing an important level of control in regulating cell-cycle progression. In mammalian cells, E2F is a family of heterodimeric transcription factors that function in the integration process. Since the original identification of E2F DNA-binding complexes [1], there has been an explosion in research activity focused on understanding E2F. A major step forward came from the discovery that the protein product of the retinoblastoma tumour suppressor gene (pRb) forms a physical complex with E2F [2]. The interaction between pRb (and the pRb-related proteins p107 and p130) and E2F can be disrupted by the action of viral oncoproteins, such as adenovirus EIA, which form a physical complex with pRb and thereby release active E2F. Furthermore, an analysis of mutant *Rb* alleles (occurring in about 20% of human tumours) showed that mutant pRb frequently cannot bind to E2F [3]. Overall, these properties underscore the important role that E2F plays in orchestrating early-cell-cycle control by allowing cell-cycle progression to be integrated with the transcription apparatus, and emphasize the considerable impact that aberrant control of E2F activity will have in maintaining the proliferation of tumour cells.

[1]*To whom correspondence should be addressed (e-mail: N.LaThangue@ bio.gla.ac.uk).*

The retinoblastoma protein family

pRb is a 110 kDa nuclear phosphoprotein and functional inactivation of both copies of the *Rb* gene is an invariant feature of both sporadic and familial retinoblastomas [4]. pRb, p107 and p130 form the pocket protein (PP) family that is crucial in cell-cycle regulation [4,5]. Certain viral oncoproteins, including adenovirus E1A and SV40 large T antigens, target the pocket domain, displacing cellular proteins and thereby leading to loss of PP function (Figure 1).

At least three distinct protein-binding activities of pRb are likely to be important for its function: the large pocket (amino acids 395–876) binds to E2F, the small pocket (amino acids 379–792) binds to LXCXE motif proteins (L, leucine; C, cysteine; E, glutamic acid; X, any amino acid) and the C pocket binds to the tyrosine protein kinase c-Abl and murine double minute clone 2 oncoprotein (MDM2) (see later) [6,7]. LXCXE motif proteins include viral

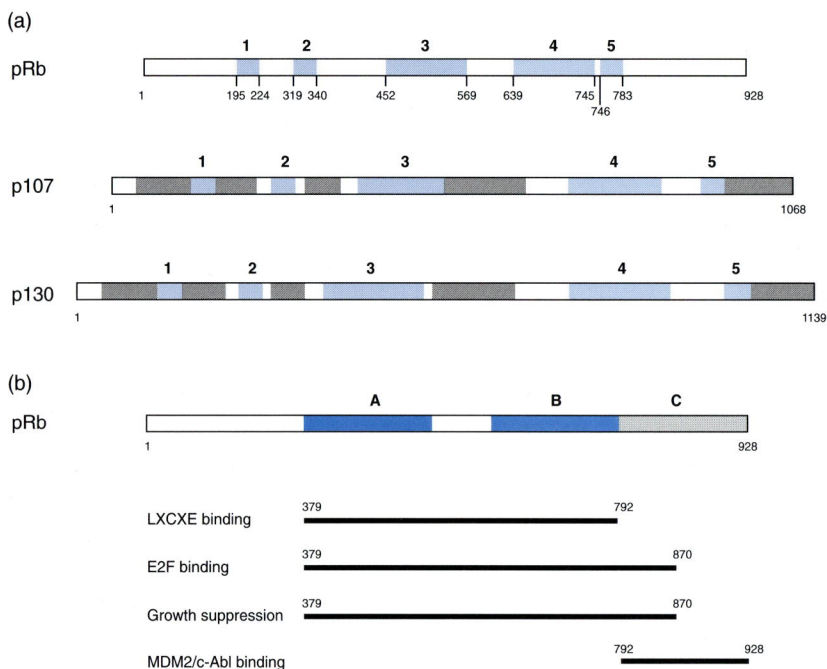

Figure 1. The PP family and functional domains in pRb
(a) Schematic organization of pRb, p107 and p130. Regions conserved among the three PPs are highlighted in blue. Grey areas indicate additional regions of similarity between p107 and p130. (b) Functional and biochemical domains (A, B, C) within pRb that are required for interaction with the LXCXE motif proteins, E2F, MDM2 and c-Abl, and for growth suppression. The A and B domains constitute the pocket domain of pRb.

oncoproteins and some cellular proteins, for example histone deacetylases (HDACs) and cyclin D.

Despite the close similarities among the family members, only *Rb* has been shown to be mutated in tumour cells. Neither *p107* nor *p130* have been found to be frequently mutated in naturally occurring tumours. Knockout mice in *Rb*, *p107* and *p130* have been developed [8,9] and their phenotypes indicate both distinct and overlapping functions among the family members. The absence of pRb causes embryonic lethality, although $p107^{-/-}$ and $p130^{-/-}$ mice survive to term, possibly as a result of functional redundancy between p107 and p130. This is consistent with the phenotype of the $p107^{-/-}:p130^{-/-}$ mice, which exhibit embryonic lethality. Interestingly, $Rb^{+/-}$ mice do not suffer from retinoblastoma, but develop tumours of the pituitary and thyroid origin, probably attributable to the different physiology of humans and mice. However, bilateral, multifocal retinal dysplasia is observed in $Rb^{+/-}:p107^{-/-}$ mice. This suggests that pRb and p107 may share overlapping functions in controlling cellular homoeostasis in the murine retina and that loss of both is required for tumour formation.

Molecular complexity of E2F

In mammalian cells E2F is a heterodimer composed of one of six E2F family members bound to one of a family of DP proteins (usually DRTF protein 1). An expanding group of genes has been identified that contain E2F binding sites [10]. In general, E2F-responsive genes encode proteins that promote proliferation and can be broadly divided into two groups: enzymes required for DNA synthesis and replication [e.g. DHFR (dihydrofolate reductase), DNA polymerase α, thymidine kinase and cell division cycle 6 kinase (CDC6)] and regulatory proteins involved in cell-cycle control (e.g. cyclins A and E and CDC2).

E2F-1, the first to be isolated, has been studied most extensively [5]. The protein contains a C-terminal transactivation domain, which also includes the protein interface recognized by pRb [11]. A DNA-binding and dimerization domain exists in the central region of the protein that contributes to sequence-specific recognition and interaction with the DP partner [11]. The N-terminal region contains binding sites for cyclin A/cyclin-dependent kinase (CDK) 2 and Skp2 (S-phase kinase associated protein), and a nuclear localization signal [12,13]. Other E2Fs share this general organization but differ in detail.

Regulation of E2F activity

Although it is clear that E2F is a sequence-specific transactivator, whose activity is modulated through association with pRb, it is becoming increasingly apparent that E2F binding sites also mediate transcriptional repression [14,15]. For example, the *dhfr* promoter contains a classical activating E2F site, whereas the B-*myb* promoter possesses a repressing site. Indeed, most E2F-

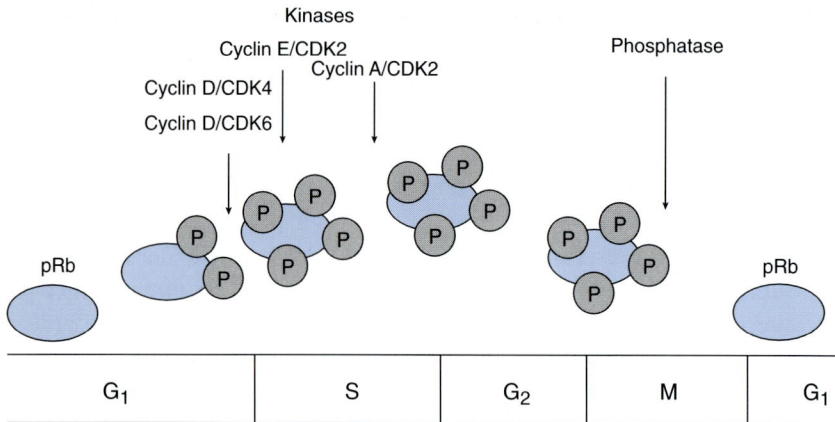

Figure 2. pRb activity is controlled by phosphorylation and dephosphorylation during the cell cycle
Hypophosphorylated pRb is active in growth repression. Phosphorylation of pRb in late G_1 by a combination of cyclin/CDKs converts pRb into an inactive hyperphosphorylated form. Dephosphorylation of pRb in late mitosis by type I protein phosphatase restores pRb growth suppressive activity.

responsive genes have been categorized as possessing activating or repressing sites. In this respect it is likely that pRb family members play an important role not only in down-regulating E2F-dependent activation by occluding the transactivation domain, but also in facilitating E2F-dependent transcriptional repression.

A simple working model for how pRb regulates E2F function suggests that during G_0/G_1, hypophosphorylated pRb binds to E2F, thus inactivating, and thereby preventing, cell-cycle progression [16]. Cyclin D/CDK4/6 and cyclin E/CDK2 progressively phosphorylate pRb in late G_1 to early S phase. In S phase, pRb phosphorylation is maintained by cyclin A/CDK2 (Figure 2). The increased phosphorylation of pRb results in reduced affinity for E2Fs, the release of free E2F and thereafter induction of E2F-responsive genes.

Different PPs have a preference for binding to different E2F family members [14,15]. E2F-1, -2 and -3 exhibit high-affinity binding towards pRb but bind weakly to p107–p130. In contrast, E2F-4 and -5 show greater specificity for p107–p130 and also bind pRb. Furthermore, complexes between the pRb family members and E2F form at different phases of the cell cycle. In general, p130–E2F is mainly found in quiescent or differentiated cells, p107–E2F predominates in S-phase cells and pRb–E2F is most evident during G_1/S phase transition, but also exists in quiescent or differentiated cells [14,15].

pRb can modulate E2F transcription activity in at least two distinct ways [14]. Firstly, pRb binds to the transcription-activation domain in E2F and directly inhibits transcriptional activity. Using *in vitro* transcription and footprinting assays, Ross et al. [17] demonstrated that pRb blocks the recruitment of the transcription initiation complex by E2F. Secondly, pRb was shown to bind to HDACs [18], which is believed to result in deacetylation of histones on the promoter and active repression of transcription via chromatin remodelling. A third mechanism is implied by the recent demonstration [19] of a pRb–HDAC–hSWI/SNF complex in which hSWI/SNF has chromatin-remodelling activity. Furthermore, it has been suggested that the pRb–hSWI/SNF complex also exists in the absence of HDAC and that this complex is sufficient to repress some genes but not others. Therefore pRb–HDAC–hSWI/SNF and pRb–hSWI/SNF could be two distinct pRb complexes that act at different phases of the cell cycle and are responsible for regulating different sets of genes during the cell cycle (Figure 3).

E2F activity may be controlled in other ways. Phosphorylation of DP-1 by cyclin A/CDK2 causes a reduction in the DNA-binding activity of the heterodimer, which may be important for down-regulating activity upon exit from S phase [20]. Phosphorylation of E2F-1 by cyclin D-dependent kinase may regulate the interaction with pRb [21]. Moreover, subunits of the heterodimer are ubiquitinylated and degraded via a proteasome-dependent pathway, a process that is likely to be important in regulating E2F during cell-cycle progression [22]. E2F activity is also regulated by nuclear accumulation and possibly export of E2F subunits, which shows a strict dependence upon the subunit composition of the heterodimer.

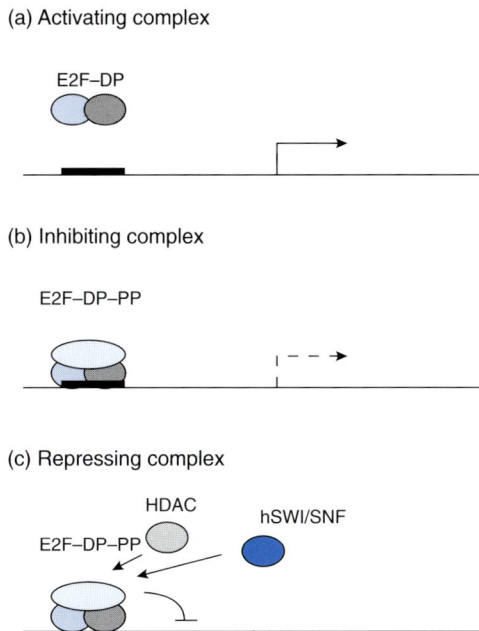

Figure 3. PPs have multiple modes of regulation in controlling E2F activity
There are at least three types of E2F complex: (a) an activating E2F complex, (b) an inhibiting E2F complex and (c) a repressing E2F complex.

Transcriptional activation by E2F

The p300/CBP [cAMP-response-element-binding protein (CREB)-binding protein] family is a group of closely related transcriptional co-activators that are believed to connect the activation domains of sequence-specific transcription factors with the transcription apparatus, thus enabling upstream signalling events to be integrated with gene expression [23]. p300/CBP proteins can physically interact with the transcriptional activation domain of E2F-1, and dominant-negative p300 mutant derivatives efficiently block E2F activation [24]. Thus it is generally believed that the p300/CBP family of co-activators plays an important role in regulating E2F transcriptional activity, although the precise mechanisms are unknown.

Control of E2F activity by phosphorylation

Recent studies have elucidated some important features in the phosphorylation control of E2F [25]. Thus E2F-5 was found to be physiologically phosphorylated by cyclin E/CDK2. Phosphorylation was found to augment the binding of the p300 co-activator, which is stabilized upon phosphorylation of the activation domain by cyclin E/CDK2 (Figure 4). Analysis of cell-cycle progression by flow cytometry established that this phosphorylation facilitates progression by promoting entry into S phase. Given that the promoter of the cyclin E gene is directly regulated by E2F, and the phosphorylation-dependent activation of E2F-5, the results imply a positive autoregulatory feedback loop that augments the interaction of E2F with p300/CBP co-activators (Figure 4).

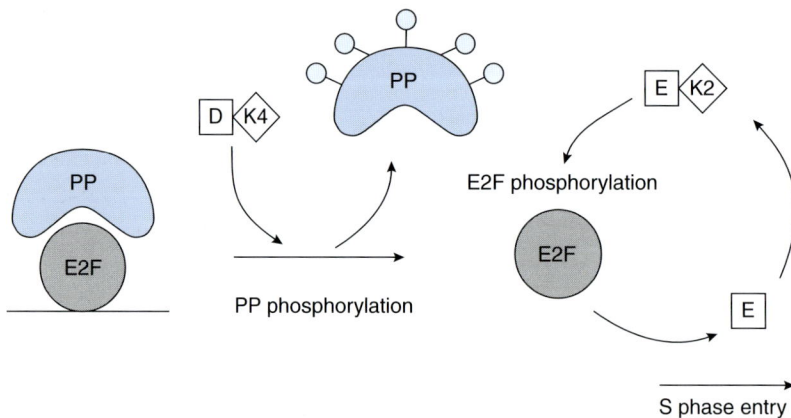

Figure 4. CDK regulation of E2F activity
It is proposed that the phosphorylation of pRb-family proteins (PPs) by the cyclin D–CDK4 complex (D–K4) releases pRb from E2F, thereafter facilitating the activation of E2F target genes. Induction of the E2F-responsive cyclin E gene allows the formation of the cyclin E–CDK2 complex (E–K2), which acts through an autoregulatory positive-feedback loop to complete the phosphorylation of pRb, initiate the phosphorylation of E2F and thereby augment E2F-dependent transcription.

Anti-apoptotic function of pRb

The over-expression of E2F-1 can cause apoptosis in SAOS2 human osteosarcoma cells [26]. Co-expression of pRb prevents this E2F-1-mediated apoptosis, which depends upon the ability of pRb to bind to E2F-1. Consistent with this, several tissues in $Rb^{-/-}$ mice display extensive cell death [8,9]. Comparison of $Rb^{-/-}$ and $E2F\text{-}1^{-/-}:Rb^{-/-}$ embryos also indicated that pRb suppresses E2F-1-mediated apoptosis.

Role of pRb in terminal differentiation

The role of pRb in differentiation was suggested from the phenotype of $Rb^{-/-}$ mice, which show a pronounced defect in erythroid, neuronal and lens development [8,9]. Although initiation of differentiation occurred, the embryos failed to achieve a fully differentiated state, indicating that pRb is likely to play an important role in regulating and maintaining the post-mitotic state. The aberrant cell-cycle entry observed in the central and peripheral nervous systems of $Rb^{-/-}$ embryos causes elevated levels of apoptosis, again implying that pRb may function in protecting cells from apoptosis.

The pRb/E2F pathway and MDM2

MDM2, the protein product of the murine double minute 2 gene (*mdm2*), is frequently over-expressed in tumour cells. The most extensively studied role for MDM2 is as a negative regulator of the p53 tumour suppressor protein, with which it forms a physical complex and thereby inactivates p53 function [27].

   Besides p53, MDM2 has been shown to interact functionally with a number of other proteins, including pRb [6] and E2F [28]. MDM2 can prevent cells from undergoing E2F-dependent apoptosis [28], an effect critically dependent on the DP component of the E2F heterodimer (Figure 5). Cells rescued from apoptosis by MDM2 possess lower levels of E2F, which appears to be dependent on the presence of the DP subunit and upon the ability of MDM2 to promote E2F degradation. Moreover, the regulation of E2F was found to correlate with an MDM2-dependent influence on the nuclear accumulation of DP-1, suggesting that an altered intracellular distribution is important in mediating the physiological effects of MDM2. Most importantly, in rescued cells MDM2 and E2F/DP can co-operate in promoting DNA synthesis and cell viability. These studies have defined a new level of interplay, between the pathways of control mediated by MDM2 and E2F, which occurs independently of p53. Furthermore, they support a model that is relevant to tumorigenesis, whereby MDM2 can antagonize the apoptotic properties of E2F and thus maintain E2F in a permanent state of growth stimulation.

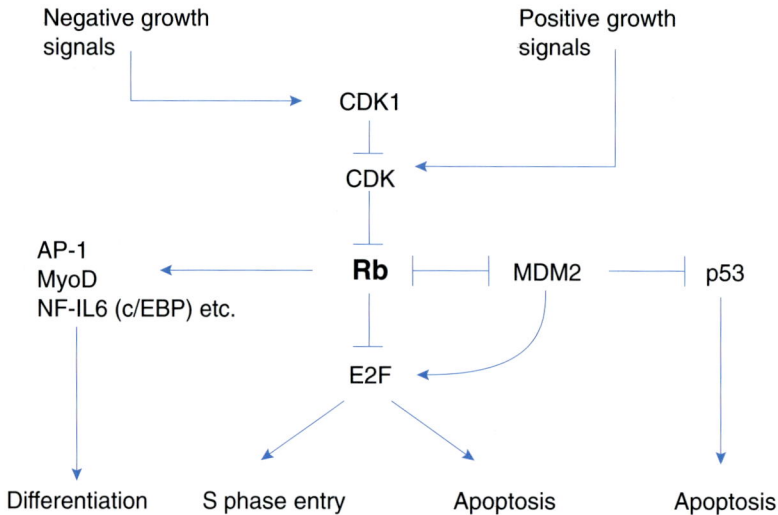

Figure 5. pRb integrates negative and positive signals with the cell-cycle clock
Mitogenic (positive growth) signals activate CDK activity and inactivate pRb function by phosphorylation. In reverse, inhibitory (negative growth) signals, such as transforming growth factor-β
(TGF-β) and contact inhibition, mobilize CDK inhibitors (CDKI) to maintain active pRb in its
hypophosphorylated form. pRb has at least two distinct mechanisms in tumour suppression:
inactivating downstream E2F activity and augmenting the activities of various transcription factors
for cellular differentiation. MDM2 acts as a mediator, bringing together the p53 and pRb pathways. AP-1, activator protein-1. MyoD and NF-IL6 (c/EBP) are transcription factors involved in
myogenesis and adipogenesis respectively.

E2F in human tumour cells

The central importance of E2F in regulating cell-cycle progression is
underscored by the variety of mutational events that occur in tumour cells
giving rise to aberrant control of E2F activity [5,14]. Although mutation of *Rb*
occurs with high frequency in a substantial proportion of human tumours,
wild-type pRb activity can be compromised either through the increased
activity of upstream CDKs that regulate Rb phosphorylation or through
mutation of CDK inhibitors. Most importantly, it is currently believed that
E2F activity is under aberrant control in most human tumour cells [4,5].

The fact that E2F and DP subunits are endowed with the ability to promote S phase progression and oncogenic activity emphasizes the potential
impact of de-regulated E2F activity on cellular physiology [9,14]. However,
although *E2F-1*$^{-/-}$ mice manifest an increase in tumour incidence, particularly
in older animals, there is good evidence that, in *Rb*$^{+/-}$ mice, tumour progression is substantially slower when E2F-1 activity is reduced.

Other targets of pRb

Low-penetrance pRb mutants, such as 661W, where tryptophan is substituted for arginine-661 in the B pocket of pRb, which are inactive in both E2F and LXCXE binding, but still retain tumour-suppressor activity. In cell-based assays, the 661W mutant was shown to inhibit G_1/S progression. Furthermore, C-pocket mutations in full-length pRb also reduce pRb function [7]. Taken together, these results suggest non-E2F targets of pRb contribute to pRb tumour suppression function. The rest of this section focuses on two examples that are involved in cell-cycle control.

The C-terminus of pRb interacts with the c-Abl tyrosine kinase [7]. This kinase is found in both the cytoplasm and nucleus. The nuclear activity is under cell-cycle control, being activated during cell-cycle progression. Interestingly, c-Abl can interact with pRb even when the pocket region is occupied by E2F.

pRb also interacts directly with $TAF_{II}250$ (the largest of the TATA-box-binding-protein-associated factors) through multiple regions in each protein [29]. Apart from being part of the basal transcriptional machinery, $TAF_{II}250$ possesses intrinsic histone acetylase activity and kinase activity. Mutagenesis studies suggest that $TAF_{II}250$ is a cell-cycle-regulated protein, and its HDAC activity is required for cell-cycle progression. pRb inhibits the kinase activity of $TAF_{II}250$ [30], suggesting that pRb regulates transcription by modulating the activity of the basal transcription apparatus.

Summary

- *The pRb tumour suppressor protein is an essential component of the cell-cycle clock, integrating both positive and negative signals for cellular growth and proliferation with the transcription machinery.*
- *pRb exerts its tumour suppression function by both antagonizing and synergizing with downstream effectors, such as E2F.*
- *pRb has two modes of action, it can inactivate E2F transcription activity or it can assemble an active repression complex with E2F.*
- *Apart from E2F, pRb interacts with various factors to promote cellular differentiation. The differentiation properties of pRb are likely to contribute partly to its tumour suppressor function.*
- *It is also clear that pRb is a master regulator for transcription. It can both activate and repress transcription in a context-dependent manner. pRb interacts directly with histone acetyltransferase, histone deacetylases and SWI/SNF proteins, all of which are classes of proteins involved in chromatin remodelling.*
- *Last, but not least, pRb regulates transcription driven by all three polymerases, thereby integrating the cell-cycle clock with the biosynthetic capacity of the cell in controlling cellular proliferation and growth.*

We thank Marie Caldwell for help in preparation of the manuscript. Work in our laboratory is supported by the Medical Research Council and the Cancer Research Campaign. H.M.C. is supported by the Wellcome Trust.

References

1. La Thangue, N.B. & Rigby, P.W.J. (1987) *Cell* **49**, 507–513
2. Bandara, L.R. & La Thangue, N.B. (1991) *Nature (London)* **351**, 494–497
3. Bandara, L.R., Adamczewski, J.P., Hunt, T. & La Thangue, N.B. (1991) *Nature (London)* **352**, 249–251
4. Riley, D.J., Lee, E.Y. & Lee, W.H. (1994) *Annu. Rev. Cell Biol.* **10**, 1–29
5. La Thangue, N.B. (1994) *Curr. Opin. Cell Biol.* **6**, 443–450
6. Xiao, Z., Chen, J., Levine, A.J., Modjtahedi, N., Xing, J., Sellers, W.R. & Livingston, D.M. (1995) *Nature (London)* **375**, 694–697
7. Wang, J.Y. (1997) *Curr. Opin. Genet. Dev.* **7**, 39–45
8. Lipinski, M.M. & Jacks, T. (1999) *Oncogene* **18**, 7873–7882
9. Mulligan, G. & Jacks, T. (1998) *Trends Genet.* **14**, 223–229
10. Farnham, P.J. (1995) *Curr. Top. Microbiol. Immunol.* **208**, 1–30
11. Bandara, L.R., Buck, V.M., Zamanian, M., Johnston, L.H. & La Thangue, N.B. (1993) *EMBO J.* **13**, 4317–4324
12. Krek, W., Livingston, D.M. & Shirodkar, S. (1993) *Science* **262**, 1557–1560
13. Krek, W., Bwen, M., Shirodkar, S., Arany, Z.Z., Kaelin, W.G. & Livingston, D. (1994) *Cell* **78**, 161–172
14. Dyson, N. (1998) *Genes Dev.* **12**, 2245–2262
15. Helin, K. (1998) *Curr. Opin. Genet. Dev.* **8**, 28–35
16. Mittnacht, S. (1998) *Curr. Opin. Genet. Dev.* **8**, 21–27
17. Ross, J.F., Liu, X. & Dynlacht, B.D. (1999) *Mol. Cell* **3**, 195–205
18. Brehm, A., Miska, E.A., McCance, D.J., Reid, J.L., Bannister, A.J. & Kouzarides, T. (1998) *Nature (London)* **391**, 597–601
19. Zhang, H., Gavin, M., Dahiya, A., Postigo, A.A., Ma, D., Luo, R.X., Harbour, J.W. & Dean, D.C. (2000) *Cell* **101**, 79–89
20. Krek, W., Xu, G. & Livingston, D.M. (1995) *Cell* **83**, 1149–1158
21. Fagan, R., Flint, K.J. & Jones, N. (1994) *Cell* **78**, 799–811
22. Marti, A., Wirbelauer, C., Scheffner, M. & Krek, W. (1999) *Nat. Cell Biol.* **1**, 14–19
23. Shikama, N., Lyon, J. & La Thangue, N.B. (1997) *Trends Cell. Biol.* **7**, 230–236
24. Lee, C.-W., Sorensen, T.S., Shikama, N. & La Thangue, N.B. (1998) *Oncogene* **16**, 2695–2710
25. Morris, L., Allen, K.E. & La Thangue (2000) *Nat. Cell. Biol.* **12**, 232–239
26. Hsieh, J.K., Fredersdorf, S., Kouzarides, T., Martin, K. & Lu, X. (1997). *Genes Dev.* **11**, 1840–1852
27. Momand, J., Zambetti, G.P., Olson, D.C., George, D.L. & Levine, A.J. (1992) *Cell* **69**, 1237–1245
28. Loughran, O. & La Thangue, N.B. (2000) *Mol. Cell. Biol.* **20**, 2186–2197
29. Shao, Z., Siegert, J.L., Ruppert, S. & Robbins, P.D. (1997) *Oncogene* **15**, 385–392
30. Siegert, J.L. & Robbins, P.D. (1999) *Mol. Cell. Biol.* **19**, 846–854

8

Regulation of mRNA translation

Christopher G. Proud[1]

Division of Molecular Physiology, School of Life Sciences, MSI/WTB Complex, University of Dundee, Dundee DD1 5EH, U.K.

Introduction

The control of mRNA translation plays a key role in regulating gene expression under a wide range of circumstances in eukaryotic cells, and to a lesser extent in bacteria. The field has grown enormously in recent years and I will therefore only attempt to give a general overview of the types of translational regulation which occur, discuss some of the situations in which it is important and outline some of the mechanisms by which it is achieved.

Why regulate mRNA translation?

Controlling translation offers advantages over other levels of control of gene expression under particular conditions, as follows.

Rapidity of response

Where cells need to increase rapidly the synthesis of particular proteins, the ability to up-regulate the translation of pre-existing mRNAs clearly allows the cell to start to make the corresponding protein without a requirement to activate transcription, process the resulting transcript and transport it to the cytoplasm. Such situations include stimulation by mitogens, growth factors and hormones, and also responses to nutrient availability and to certain stressful conditions. Examples of these kinds of control are discussed in detail later.

[1]*e-mail: c.g.proud@dundee.ac.uk*

Independence from transcription

The ability to control translation is essential where there is little or no ongoing transcription, for example, in early development of a number of organisms, where transcription is essentially switched off, and in cells which have no functional nucleus. One example of the latter is the mammalian reticulocyte. Globin production is regulated by haemin, but since reticulocytes lack a functional nucleus, regulation of transcription is not an option and translation is the control point.

Localized protein expression

If a protein is only required at a specific location in the cell, this can be achieved in various ways. Firstly, the mRNA can be localized within the cell. Secondly, the protein can be made throughout the cell and then transported to its destination. A third option is to localize the translation of the mRNA, which may itself be present widely in the cell. Important examples of the first and third options play key roles in controlling gene expression during development, e.g. to generate intracellular gradients of proteins involved in specifying cell polarity.

Protein synthesis as an anabolic process

Activation of protein synthesis is also an important response to stimuli which regulate cell growth. For example, insulin promotes protein synthesis and anabolic processes such as glycogen storage in a variety of tissues and cell types. The effect of such stimuli on protein metabolism is, of course, a balance between its effects on protein synthesis and degradation.

The overall process of mRNA translation

Translation of a typical eukaryotic mRNA is summarized in Figure 1. The 5′-

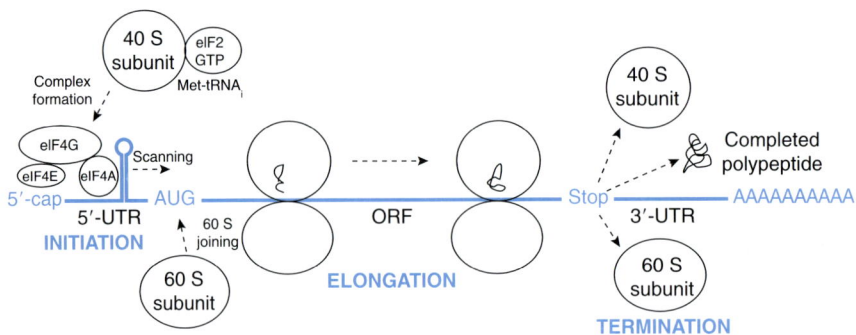

Figure 1. The overall process of mRNA translation
The figure schematically depicts an mRNA undergoing translation as a thick straight line. However, it does not show the communication between the 5′- and 3′-ends which we now know occurs. ORF, open reading frame.

untranslated region (5′-UTR, also called the 5′-leader) of an mRNA may contain features important in regulating its translation; for example, stable stem–loops in the 5′-UTR are inhibitory. Sequence elements in the 3′-UTR also often play key roles in regulating mRNA translation or stability, and the length of the polyadenylated [poly(A)] tail is also important in regulatory translation.

Initiation of translation is generally considered to be of prime importance in regulating translation. Eukaryotic initiation factor 4E (eIF4E) binds to the 5′-cap structure and forms a complex with the scaffolding protein eIF4G [1,2]. eIF4A can unwind double-stranded (ds)RNA and binds to eIF4G/4E to form eIF4F. As part of eIF4F, eIF4A efficiently unwinds secondary structures and thus facilitates mRNA translation. The 40 S subunit binds a ternary complex of eIF2, GTP and the initiator methionyl-tRNA (Met-tRNA$_i$) to form the 43 S initiation complex. The anti-codon on the Met-tRNA$_i$ recognizes the start codon (usually the first AUG in the mRNA). Although the precise details of the process remain to be established, it is now generally accepted that the 43 S complex (perhaps in association with other eIFs) 'scans' along the mRNA, starting at the 5′-end, to locate the start codon. Stem–loops in the 5′-UTR thus probably interfere with translation by impairing the movement of the scanning initiation complex. Some mRNAs utilize a quite different 'internal ribosome entry' mechanism (see later).

Examples of specific types of translational control

Regulation of mRNA translation by nutrients

In *Saccharomyces cerevisiae*, amino acid starvation leads to increased expression of the genes that encode enzymes involved in amino-acid biosynthesis. This effect involves a dramatic increase in the translation of the mRNA for a transcriptional activator protein, Gcn4p, and requires the product of the *GCN2* gene, the protein kinase Gcn2p [3]. Gcn2p phosphorylates the smallest subunit (α) of eIF2 at a single serine residue in a sequence that is highly conserved among eukaryotic organisms (Figure 2a). eIF2(αP) is a potent competitive inhibitor of eIF2B, a multi-subunit protein that regenerates active eIF2·GTP from inactive eIF2·GDP by a process of nucleotide exchange. Therefore, phosphorylation of eIF2 reduces the amount of active eIF2 in the cell and impairs the ability of the 40 S subunit to acquire Met-tRNA$_i$. This underlies the 'trick' by which an effect that impairs overall initiation actually brings about an increase in the translation of the mRNA for Gcn4p. Its 5′-leader contains four upstream open reading frames (uORFs), i.e. short coding sequences each equipped with their own start and stop codons. The 43 S complexes are thought to recognize the initial AUG in the GCN4 mRNA, leading to translation of uORF1 and termination at its stop codon, as expected. It is believed that the ribosomal subunits remain associated with the mRNA and recommence scanning. This requires eIF2·GTP·Met-tRNA$_i$. If

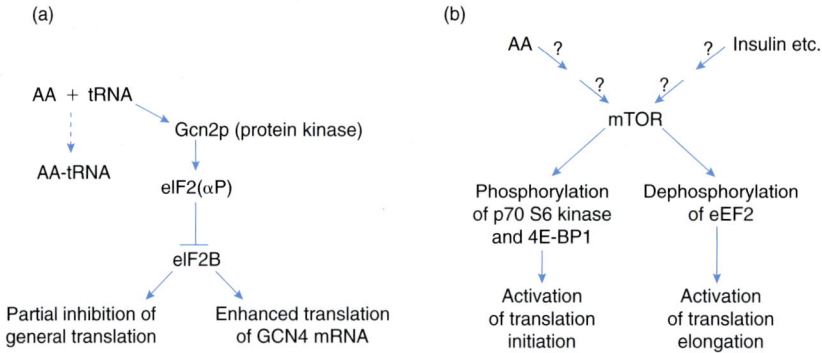

Figure 2. Regulation of mRNA translation by nutrients
(a) Regulation by amino acids in yeast via the eIF2α protein kinase, Gcn2p. (b) Regulation by amino acids in mammalian cells through mTOR-linked signalling pathways. Solid arrows indicate regulatory or signalling connections. The dashed arrow indicates the aminoacylation of tRNA. ⊥ indicates inhibition. eEF2 is an elongation factor. AA, amino acid; 4E-BP, eIF4E-binding protein.

eIF2 activity is high, ribosomes will quickly reacquire Met-tRNA$_i$ and recognize the AUG of a subsequent uORF. However, once this uORF has been translated, the ribosomes leave the mRNA and therefore do not reach the GCN4 ORF which lies downstream. Gcn4p is therefore not produced. If, however, eIF2 activity is low, due to activation of Gcn2p and phosphorylation of eIF2α, for example, ribosomes will not readily acquire Met-tRNA$_i$ and some will move beyond uORF4 before doing so. They will therefore remain bound to the mRNA. If they subsequently bind eIF2 and Met-tRNA$_i$, they are able to recognize the start of the main GCN4 ORF and translate it. Thus reducing eIF2 activity has the apparently paradoxical effect of increasing the translation of the downstream GCN4 ORF.

Many other mRNAs also possess uORFs, although these are frequently single uORFs rather than the multiple ones found in the GCN4 mRNA. None has been studied in as much detail and no other mRNA has yet been convincingly shown to be regulated in a similar way. Since animals possess Gcn2 homologues, and other eIF2 kinases, it seems likely that similar mechanisms await discovery in higher organisms. There are mammalian examples where uORFs modulate translation of a downstream ORF, but not in the same way as described above for GCN4 [4].

A further mechanism by which nutrients regulate translation initiation involves the proteins known as TOR (in yeast) or mTOR (in mammals) (Figure 2b) [5]. TOR stands for target-of-rapamycin, rapamycin being an immunosuppressant compound which, when complexed with a binding protein, inhibits the (unknown) function of (m)TOR. In both mammals and yeast, the TOR proteins are involved in signalling events that regulate mRNA translation and respond to nutrient availability. In yeast, rapamycin causes profound inhibition of translation initiation. In mammals, mTOR regulates sever-

al proteins involved in mRNA translation, including the eIF4E-binding proteins (4E-BPs), whose binding to eIF4E blocks formation of eIF4F [6]. Insulin
and other agents bring about the phosphorylation of 4E-BPs, allowing them to
dissociate from eIF4E, thereby facilitating eIF4F formation [6,7]. In several
cell types, these effects require amino acids to be present in the medium.
Amino acids also serve to maintain the basal level of 4E-BP phosphorylation
and thus exert a form of 'feed-forward' activation of eIF4F formation, which
makes excellent physiological sense as amino acids are the precursors for protein production.

Rapid regulation of the translation of specific mRNAs

The translation of many mRNAs is subject to short-term regulation by
hormones or environmental conditions. One group is the 5'-TOP mRNAs,
where TOP stands for 'tract of pyrimidines' to denote that they contain a
sequence of such nucleotides at their extreme 5'-end (Figure 3a) [8]. These
mRNAs generally encode ribosomal proteins and other components of the
translational machinery. In serum-deprived cells, 5'-TOP mRNAs are poorly
translated. Treatment of the cells with serum causes them to bind ribosomes
and shift into polyribosomes. How does serum bring about the activation of
their translation? The effect is blocked by rapamycin, indicating an
involvement of mTOR. The available evidence suggests that the mechanism

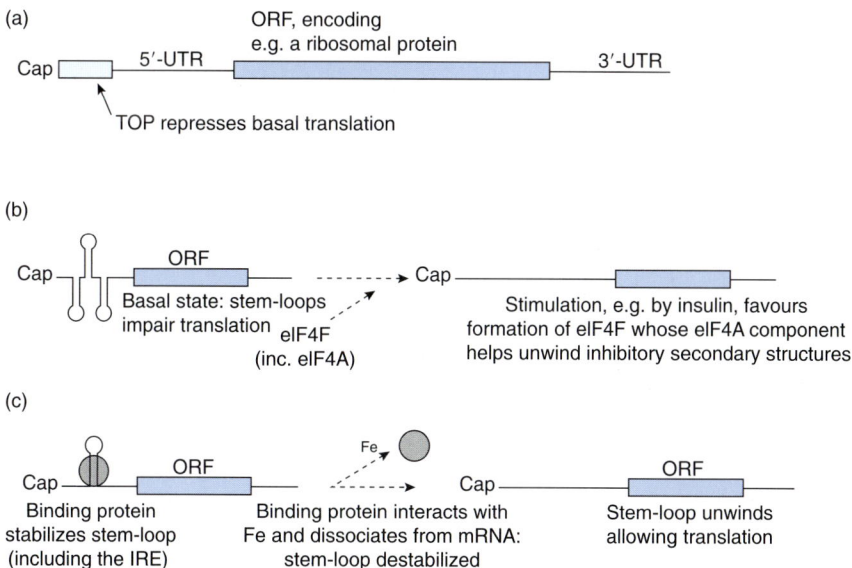

Figure 3. Themes and variations in translational control
Elements in the 5'-UTR can regulate the translation of specific mRNAs. (a) 5'-TOP mRNAs. (b)
Structured 5'-UTR. (c) Ferritin. See text for details of each example. All diagrams are schematic;
no particular structure or scale is implied.

involves a protein kinase that phosphorylates the 40 S subunit protein S6, termed p70 S6 kinase, which is regulated via the mTOR pathway (Figure 2b). S6 is located close to the mRNA binding site in the 40 S subunit, and its phosphorylation may facilitate the binding or translation of the 5′-TOP mRNAs. Since mTOR function is apparently regulated by amino acid availability, this mechanism allows amino acids to positively regulate ribosome synthesis, which makes excellent sense in terms of the cellular economy.

The mRNAs for a number of growth factors, transcription factors and other proteins linked to cell proliferation are predicted to contain stable stem-loops within their 5′-UTRs (Figure 3b) [9,10]. Such mRNAs are likely to be regulated by the availability of the eIF4F complex, which contains the helicase eIF4A. Indeed, the translation of mRNAs with structured 5′-UTRs is enhanced by overexpression of eIF4E, which should favour eIF4F formation. Particularly exciting findings in relation to this are that eIF4E overexpression can lead to malignant transformation of cells, and eIF4E levels are high in some cancer cells, especially in more 'aggressive' tumours. Elevated levels of eIF4E may facilitate expression of growth-associated proteins, contributing to the aberrant growth properties of tumour cells.

Ferritin is required for storage of iron. It is important to capture iron when available, not only because it is essential for the production of haemo-proteins, such as haemoglobin and cytochromes, but also because iron ions are toxic. Ferritin synthesis is rapidly enhanced in response to increased availability of iron, without a corresponding change in mRNA levels, indicating that iron enhances ferritin mRNA translation [11]. Ferritin mRNAs contain a highly conserved, but not very stable, stem–loop within their 5′-UTRs (Figure 3c), termed the iron-response element (IRE). Elegant experiments have demonstrated that the IRE inhibits basal mRNA translation and confers sensitivity to iron. It does so by interacting with a protein, the IRE-binding protein, which stabilizes the loop and thus causes it to inhibit translation (presumably by blocking scanning; see above). Upon binding iron, the IRE-binding protein loses its ability to bind to the mRNA, rendering the stem–loop less stable and facilitating efficient translation of the mRNA. Iron thus promotes the synthesis of the protein which stores it!

Translation in virus-infected cells

Viruses do not possess their own translational machinery and must therefore utilize that of the host cell in order to produce the proteins that they encode. Viruses must ensure that their mRNAs are translated efficiently and, in certain cases, shut off host-cell protein synthesis, to maximize translation of their own mRNAs. An excellent example of both phenomena is provided by the *Picornaviridae*, RNA viruses which possess one mRNA encoding all the polypeptides required for viral replication/production as a polyprotein [12]. This is later cleaved into various smaller peptides. Examples include poliovirus (PV), foot and mouth disease virus and hepatitis C virus. Picornaviral mRNAs

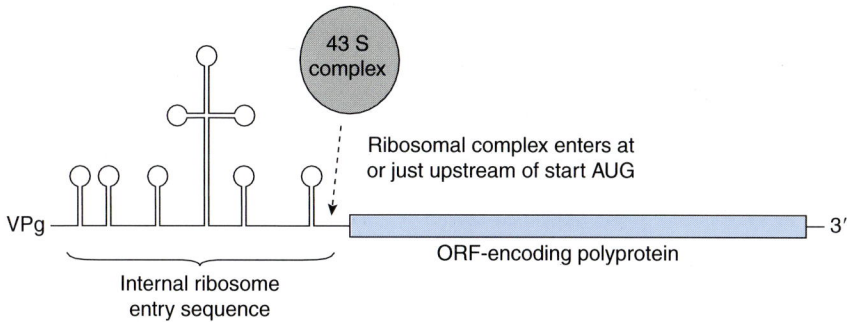

Figure 4. Internal ribosome entry permits efficient translation of PV mRNA despite host cell translation being shut off
See text for further explanation. VPg indicates the virus-encoded protein that is covalently linked to the 5'-terminus of picornaviral mRNAs.

are not capped, but instead have a protein at their 5'-end and possess long and highly structured 5'-UTRs (Figure 4). These features suggest that their translation must occur by a mechanism different from normal cap-dependent scanning. It is now well established that their 5'-UTRs contain a so-called internal ribosome entry site, which allows 40 S subunits to bind to the mRNA such that they are positioned at or near the start codon without scanning. In contrast, translation of cellular messages is severely inhibited in infected cells. Why? PV infection results in cleavage of eIF4G, such that the region which interacts with eIF4E is separated from that which binds eIF3 and eIF4A. This results in inhibition of translation of capped mRNAs, which require eIF4E to form eIF4F complexes, but not of PV mRNA, which, being uncapped, does not. Thus, at a stroke, PV accomplishes both its aims — inhibition of host translation *and* facilitation of the translation of its own mRNA. Such internal ribosome entry site elements are present not only in the mRNAs for other picornaviruses, but also in several cellular mRNAs. This should allow such mRNAs to continue to be translated in situations where cap-dependent translation is inhibited, such as conditions of nutrient or serum starvation, or certain stress conditions.

It is clearly important for host cells to try to prevent viral replication. One way in which they do this involves the dsRNA-activated protein kinase, PKR, which phosphorylates eIF2α at the same site as GCN2, thus inhibiting eIF2B and translation initiation [13,14]. PKR is induced by interferon as part of the anti-viral response, but only becomes active in the presence of dsRNA, generated during virus infection. PKR activation inhibits protein synthesis, thus preventing viral replication. However, viruses fight back. Some block activation of PKR by producing proteins that bind dsRNA (vaccinia virus), or dsRNA molecules that bind to but do not activate PKR (adenovirus). Others produce or activate proteins which block PKR activity (influenza, vaccinia). Herpes simplex virus activates the phosphatase that dephosphorylates eIF2α and opposes the action of PKR. The fact that viruses possess such varied ways of tackling PKR testifies to its importance in blocking viral replication.

Overall activation of protein synthesis

So far we have dealt mainly with mechanisms by which the translation of specific mRNAs, or subsets of mRNAs, are regulated. How about overall protein synthesis? It is increased under a variety of conditions, e.g. by insulin and by other growth-promoting stimuli such as mitogens and growth factors. Cell division requires cells to attain a certain critical size (enough to form two daughter cells) and thus it is important that proliferative agents activate translation. Two translation factors required for the translation of all mRNAs are activated acutely, e.g. by insulin. These are eIF2B (which recycles eIF2, see above) and the eukaryotic elongation factor 2 (eEF2) [7]. Both are inhibited by phosphorylation at specific sites which undergo dephosphorylation in response to insulin.

As shown in Figure 5(a), eIF2B is phosphorylated and inhibited by glycogen synthase kinase-3 (GSK-3), which is inactivated by insulin through a signalling pathway involving phosphoinositide 3-kinase and probably protein kinase B, which phosphorylates and inactivates GSK-3. This results in dephosphorylation and activation of eIF2B.

eEF2 is phosphorylated and inactivated by a specific protein kinase, eEF2 kinase [15]. eEF2 also undergoes rapid dephosphorylation in response to

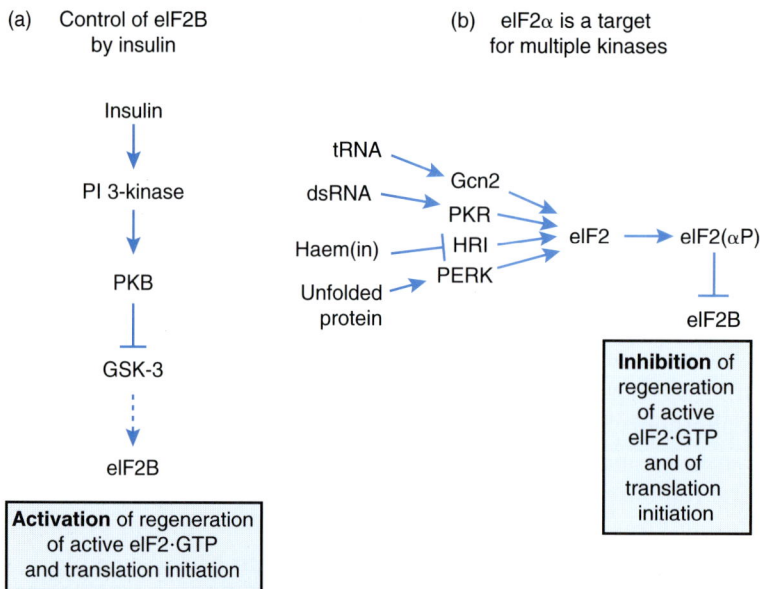

Figure 5. Overall control of protein synthesis
(a) Regulatory events impinging upon eIF2B. (b) Four related protein kinases phosphorylate eIF2α. Abbreviations: PI 3-kinase, phosphoinositide 3-kinase; PKB, protein kinase B; HRI, haem-regulated inhibitor.

insulin; this correlates with the activation of elongation. Both effects are blocked by rapamycin, identifying eEF2 as a third target in translation for mTOR signalling and for regulation by amino acids as well as hormonal/mitogenic signals (Figure 2b). eEF2 phosphorylation is increased by events which raise the levels of two important second messengers, Ca^{2+} and cAMP, which activate eEF2 kinase either directly or indirectly. By inhibiting protein synthesis, these effects may help conserve ATP for other cellular processes, e.g. in skeletal or cardiac muscle where these second messengers are important in regulating contraction, a process of paramount importance for the survival of the animal and one which requires considerable metabolic energy (about 20% of total cellular ATP consumption).

Certain specific conditions also lead to inhibition of overall protein synthesis. Two such examples involve phosphorylation of eIF2α (Figure 5b). In the absence of haem, reticulocyte protein synthesis is shut off, preventing wasteful production of unwanted globin. Shut-off is due to activation of an eIF2α kinase, the haem-controlled repressor HRI (haem-regulated inhibitor) [2], which is inhibited by haem(in) and thus serves to link its availability to globin production. The second example involves the fourth known eIF2α kinase, which is resident in the endoplasmic reticulum (ER) membrane (PERK, for the PKR-like ER-resident kinase) [2,16]. This kinase is believed to link protein folding/processing in the lumen of the ER to translation in the cytoplasm. Thus if the ER is overloaded and cannot keep pace with the arrival of proteins synthesized in the cytoplasm, PERK is activated. Its cytosolic domain phosphorylates eIF2α, causing a pause in protein synthesis. Presumably, once the ER has caught up or recovered, PERK is switched off, allowing cytosolic translation to resume. The four known eIF2α kinases are thus all involved in switching off translation in responses to diverse stresses.

Localized mRNA translation during development

During early development, embryos must begin to define crucial developmental features such as their anterior and posterior poles. The regulation of the translation of certain maternal mRNAs is known to play a key role in this process, which has been studied most intensively in the fruit fly, *Drosophila melanogaster* [17]. Nanos protein is required for specification of the anterior–posterior axis in *Drosophila* embryos. Prior to fertilization of the egg, Nanos mRNA is present but is not translated. Translation of Nanos requires another protein, Oskar, which both localizes the Nanos mRNA to the posterior pole and permits its localized translation (Figure 6). Oskar protein is made prior to fertilization and this depends on localization of Oskar mRNA to the posterior pole. Nanos regulates the translation of another mRNA (for Hunchback) by binding, together with a further protein Pumilio, to sequences in the 3'-UTR of the Hunchback mRNA and repressing its translation. Since levels of Nanos protein are higher towards the posterior pole of the embryo, Hunchback translation occurs primarily towards the anterior pole of the

mRNA   Protein

1 Oskar (early)   A   P   A   P

2 Oskar (late)

3 Nanos

4 Hunchback

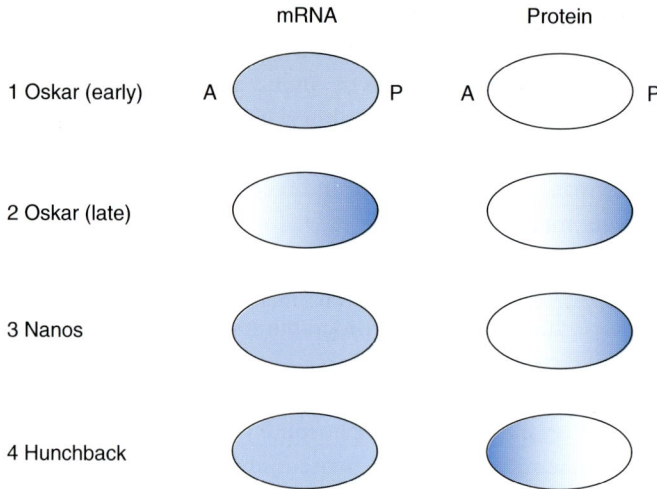

Figure 6. Translational control during early development: localized translation of specific mRNAs creates concentration gradients of proteins involved in determining cell polarity

The process is discussed in detail in the text. Intensity of shading indicates the absence or presence of a concentration gradient and its polarity. A, anterior pole; P, posterior pole.

developing embryo. This leads to the accumulation of Nanos and Hunchback proteins in opposing concentration gradients (Figure 6), with Hunchback protein accumulating at the anterior end of the embryo, where it represses transcription of abdomen-specific genes. These translation control mechanisms underlie the acquisition of anterior–posterior polarity. Their importance for normal development of the embryo is underlined by the observation that disruption of the control of Nanos translation, for example, results in abnormal patterning of the embryo, resulting, in the most severe cases, in the formation of two complete posterior abdomens in mirror-image with each other [17].

Other types of translational control during development

Two further ways in which the translation of specific mRNAs is regulated during development involve unmasking and cytoplasmic polyadenylation [17]. Unmasking refers to the fact that some mRNAs, whose translation is turned on at specific stages during gametogenesis or embryonic development, appear to be sequestered as inactive messenger ribonucleoprotein complexes at earlier times. In other words, their translation is repressed by binding to proteins which dissociate at the appropriate time, thus allowing the mRNA to be translated. How is the association of these proteins with their mRNA partners regulated? Little is known about this beyond the fact that several such proteins are phosphoproteins suggesting that changes in their phosphorylation states may regulate mRNA binding.

It is widely accepted that the 3′-poly(A) tail plays a key role in controlling the translation of certain mRNAs during gametogenesis and embryogenesis. Increased poly(A) tail length favours translation. The tails of a number of mRNAs are lengthened when they become translationally active, and when mRNAs with different tail lengths were microinjected into cells, those with long tails were found to be translated better. Tail length is regulated through the opposing actions of adenylating and deadenylating enzymes. How this is controlled is unclear, but sequences in the 3′-UTR are important in modulating tail length. How do longer tails facilitate mRNA translation? Recent data show that the 3′- and 5′-ends of mRNAs are brought together by the interaction of the poly(A)-binding protein (PABP) with eIF4F [18]. This may explain how tail length (and indeed proteins bound to the 3-UTR) can influence the translation initiation events occurring at the 5′ end of the mRNA.

Perspectives

It is now clear that many mechanisms exist to regulate mRNA translation in eukaryotic organisms, and that control of mRNA translation plays an important role in regulating gene expression in response to varied stimuli and in diverse physiological situations. These mechanisms involve control of translation factors or components of the ribosome, or of proteins which interact with the 5′- or 3′-UTRs of the mRNA. A common feature of these mechanisms is that many, if not most, involve protein phosphorylation/dephosphorylation. Many aspects still require clarification. How do the proteins which interact with the 3′-UTR exert their varied effects upon translation and poly(A) tail length (and on also mRNA stability, not discussed here)? How do the protein factors and RNA molecules really function at the molecular level in the process of mRNA translation and its control? An understanding of this requires knowledge of their three-dimensional structures and it is thus highly significant that the last 2–3 years have seen the determination of the structures of several translation factors. The signalling pathways linking the control of translation factors to cellular cues also remain to be elucidated in many cases, especially for the regulation by nutrients such as amino acids. Thirdly, there are undoubtedly many more mRNA-binding proteins awaiting discovery, for example those that interact with the 5′-UTR.

Summary

- *The control of mRNA translation plays an important role in regulating gene expression in diverse situations.*
- *Nutrients, especially amino acids, regulate translation to control the expression of specific proteins including transcriptional activators and ribosomal proteins.*

- *Sequence elements in the 5'- or 3'-untranslated regions of mRNAs can control their translation.*
- *Hormones such as insulin activate protein synthesis by switching on translation factors required for overall mRNA translation.*
- *Viruses employ a range of stratagems to ensure efficient translation of their mRNAs and in some cases to inhibit host cell translation.*
- *During development, control of mRNA translation regulates gene expression both spatially and temporally.*

References

1.   Gingras, A.-C., Raught, B. & Sonenberg, N. (1999) eIF4 translation factors: effectors of mRNA recruitment to ribosomes and regulators of translation. *Annu. Rev. Biochem.* **68**, 913–963
2.   Dever, T.E. (1999) Translation initiation: adept at adapting. *Trends Biochem. Sci.* **24**, 398–403
3.   Hinnebusch, A.G. (1994) Translational control of *GCN4*: an *in vivo* barometer of initiation-factor activity. *Trends Biochem. Sci.* **19**, 409–414
4.   Geballe, A.P. & Morris, D.R. (1994) Initiation codons within 5'-leaders of mRNAs as regulators of translation. *Trends Biochem. Sci.* **19**, 159–164
5.   Dennis, P.B., Fumagalli, S. & Thomas, G. (1999) Target of rapamycin (TOR): balancing the opposing forces of protein synthesis and degradation. *Curr. Opin. Genet. Dev.* **9**, 49–54
6.   Lawrence, J.C. & Abraham, R.T. (1997) PHAS/4E-BPs as regulators of mRNA translation and cell proliferation. *Trends Biochem. Sci.* **22**, 345–349
7.   Proud, C.G. & Denton, R.M. (1997) Molecular mechanisms for the activation of protein synthesis by insulin. *Biochem. J.* **328**, 329–341
8.   Meyuhas, O. & Hornstein, E. (2000) Translational control of TOP mRNAs, in *Translational Control of Gene Expression* (Sonenberg, N., Hershey, J.W.B. & Mathews, M.B., eds.), pp. 671–693, Cold Spring Harbor Laboratory Press, Cold Spring Harbor, NY
9.   Clemens, M.J. & Bommer, U.-A. (1999) Translational control: the cancer connection. *Int. J. Biochem. Cell Biol.* **31**, 1–23
10. Flynn, A. & Proud, C.G. (1996) The role of eIF4 in cell proliferation. *Cancer Surveys* **27**, 293–310
11. Rouault, T.A. & Harford, J.B. (2000) Translational control of ferritin synthesis, in *Translational Control of Gene Expression* (Sonenberg, N., Hershey, J.W.B., & Mathews, M.B., eds.), pp. 655–670, Cold Spring Harbor Laboratory Press, Cold Spring Harbor, NY
12. Belsham, G.J. and Jackson, R.J. (2000) Translation initiation on picornavirus RNA, in *Translational Control of Gene Expression* (Sonenberg, N., Hershey, J.W.B., & Mathews, M.B., eds.), pp. 869–900, Cold Spring Harbor Laboratory Press, Cold Spring Harbor, NY
13. Clemens, M.J. (1997) PKR - a protein kinase regulated by double-stranded RNA. *Int. J. Biochem. Cell Biol.* **29**, 945–949
14. Proud, C.G. (1996) p70 S6 kinase: an enigma with variations. *Trends Biochem. Sci.* **21**, 181–185
15. Ryazanov, A.G., Pavur, K.S. & Dorovkov, M.V. (1999) Alpha kinases: a new class of protein kinases with a novel catalytic domain. *Curr. Biol.* **9**, R43–R45
16. Harding, H.P., Zhang, Y. & Ron, D. (1999) Protein translation and folding are coupled by an endoplasmic-reticulum resident kinase. *Nature (London)* **397**, 271–274
17. Curtis, D., Lehmann, R. & Zamore, P.D. (1995) Translational regulation in development. *Cell* **81**, 171–178
18. Wells, S.E., Hillner, P.E., Vale, R.D. & Sachs, A.B. (1998) Circularization of mRNA by Eukaryotic Translation Initiation Factors. *Mol. Cell* **2**, 135–140

9

To live or die — a cell's choice

Martyn Link and David J. Harrison[1]

Department of Pathology, The University of Edinburgh, Medical School, Teviot Place, Edinburgh EH8 9AG, U.K.

Introduction

Living cells face a constant bombardment from their environment. Neighbouring cells, hormonal signals, foreign bodies and nutrients all contribute to the microcosm surrounding the cell, each interacting with the cell surface and many initiating intracellular pathways. The thousands of different proteins expressed produce hundreds of pathways with positive and negative feedback loops. In order for a multicellular organism to function correctly, each cell must efficiently decode, filter and respond to these messages. The correct response is essential to maintain the viability of the whole organism. The accumulation of detrimental signals can lead to cellular injury, after which a cell has four options (Figure 1). These options, which may not be exclusive of one another, are cell proliferation, differentiation, cell-cycle arrest (called senescence if the arrest is permanent) and genetically regulated cell death by apoptosis. (Another form of cell death, necrosis, is considered accidental and not an active choice by the cell.)

In the past few years our knowledge of the above pathways has increased dramatically. This chapter does not discuss these pathways in detail but focuses on what causes a cell to choose a particular pathway, especially apoptosis.

Cell death

Early work proposed a model of cell death that sorted the events into three stages: cell commitment, execution and clearance. In this chapter attention will be given to the first stage only — commitment. This is the crossroads at which

[1]*To whom correspondence should be addressed.*

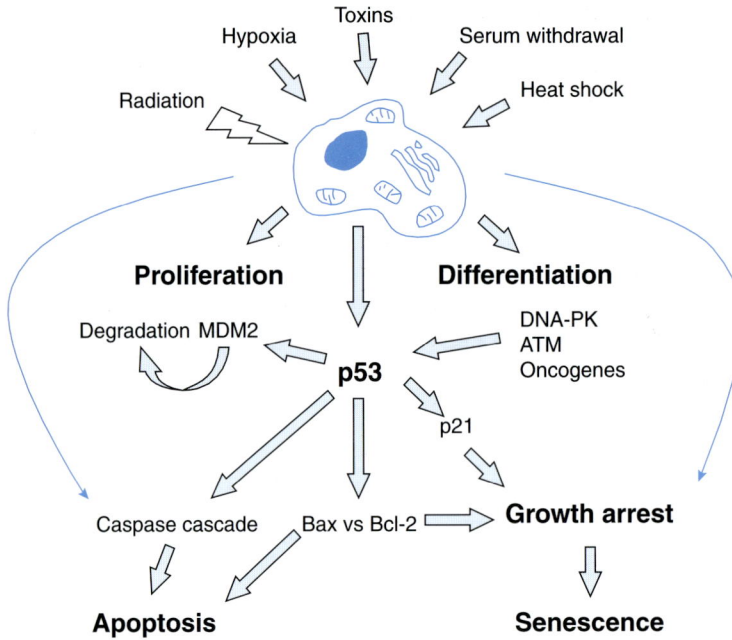

Figure 1. p53-dependent and -independent responses to injury
A wide variety of stimuli lead to p53 stabilization. Downstream of p53 stabilization there are a number of possible pathways, depending on the presence of other control genes. Other stimuli are p53-independent, acting on p53 downstream targets or on other control proteins to induce their effects. MDM2, murine double minute clone 2 oncoprotein; DNA-PK, DNA-dependent protein kinase; ATM, ataxia-telangiectasia mutated.

the cell integrates incoming signals and has to decide what action to take: whether to continue to live and divide, or press the self-destruct button.

After integration into the cell's vast communication network, most signals will effect only minor alterations in cellular function. Very little may appear to have changed from the outside and it is assumed that life goes on as normal. However, subtle changes may have occurred that are setting in motion the eventual death of the cell.

Let's start with a nematode worm, *Caenorhabditis elegans* to be precise. Several genes were identified that are necessary for, and sufficient to cause, cell death. They were named *ced-3* and *ced-4* (ced stands for *Caenorhabditis elegans* death). Another gene, *ced-9*, was found to inhibit *ced-3* and *ced-4* and is thus able to sustain life [1].

Interestingly, Bcl-2, the human homologue of the nematode CED-9 protein, is partially able to prevent cell death in *C. elegans*, revealing the level of conservation of the function of the molecule between species [2].

Inherent properties of the cell

This same stimulus can produce radically different effects in different cells. For example, thymocytes and lymphoid or myeloid cell lines undergo cell death in response to DNA-damaging ionizing radiation. However, following the same stimulus, fibroblasts respond with cell-cycle arrest [3].

Cellular context and genetic make-up

The cellular context of a cell describes the particular genes expressed in the cell and their relative protein levels. Together, these determine cellular function and how it responds to various stimuli. Cellular context is affected by the function of the cell in the body, the stage of differentiation of the cell and the intracellular micro-environment.

In the case of cell fate, the most commonly studied cells are lymphoid, myeloid, thymocyte and fibroblast cell types. Following DNA injury these cells all express p53-induced proteins such as Bax and Bcl-X_L, as well as proteins whose expression is independent of p53, including Mcl-1, GADD34 and c-Jun. Even within particular cell types there are differences in response. For example, if death-inducing surface receptors on T-lymphoid cell lines are activated, there is an increase in c-Myc expression and apoptosis results. However, B-cell lymphomas may be protected from apoptosis by c-Myc expression [4].

So which genes decide cellular fate?

Transcription factors — p53

Inducible transcription factors such as p53, E2F and c-Myc are vital for processing and amplifying cellular information. They decide cell fate as they interact with each other and recruit other genes by increasing their transcription. The loss of a gene that enables cell survival, or likewise one that prevents cell death, is likely to have disastrous consequences. In approximately half of all human tumours, p53 is mutated. This shows the importance of this gene in preventing disadvantageous proliferation.

p53 is a nuclear phosphoprotein that has been shown to be activated in response to a single-strand break in DNA, hypoxia, oxidative damage and nucleotide or oncogene imbalance [5]. Hypothesized to bind directly to sites of DNA damage or mismatches, it may serve as a damage detector itself or as part of a larger recognition complex that includes the general transcription factor, TFIIH. Importantly, p53 expression and activity are modulated at many levels within the cell. The main control areas are transcription, translation, post-translation (including phosphorylation and acetylation), cellular localization and protein stability [6] (see also Chapter 7 in this volume). Under normal conditions p53 has a short half-life, being rapidly degraded via an autoregulatory ubiquitin-dependent proteasome pathway promoted by the murine double minute clone 2 oncoprotein (MDM2).

When activated, p53 can induce either growth arrest or apoptosis (Figure 1). Both end points can be observed in different cells at different times, so what determines which pathway is chosen? One theory is that p53 is differentially transactivated; in other words, upstream regulators alter the specificity of p53 target genes. Under conditions of little DNA damage, there is only a slight increase in the level of p53 expression and growth arrest is initiated to allow time for repair of the DNA before subsequent proliferation. In the presence of large amounts of DNA damage p53 is highly expressed and cell death becomes inevitable. Another theory is that altering the cellular location of p53 determines the outcome, as it is translocated during degradation. Indeed, altering the expression levels of MDM2 would provide another means of control.

Cell cycle and the proto-oncogenes — E2F and c-Myc

The stage of the cell cycle can predetermine a cell's response to transcription factors. According to Hengstschlager et al. [7], Rat-1 cells in S phase are resistant to cell death induced by activated E2F or c-Myc. This resistance is specific, as the S phase cells were not resistant to death induced by treatment with actinomycin D. As yet the protective mechanism of S phase cells is unknown. Cells in G_1 and G_2 were susceptible to death from c-Myc expression, whereas E2F caused death specifically in G_1. This reveals the importance of the cell cycle in determining cell fate.

E2F is a growth-promoting transcription factor. In the E2F family there are five subsets — E2F-1–5. During times of cellular rest, E2F is bound by one of the retinoblastoma (Rb) family of phosphoproteins. E2F-1, E2F-2 and E2F-3 bind to Rb-1, E2F-4 to p107, and E2F-5 to p130 (see Chapter 7 by H.M. Chan et al.). In its active, unbound form E2F will drive the cell cycle through G_1 to S phase and DNA replication. By maintaining a high proportion of bound E2F, the cell can control cell-cycle progression.

Shan and Lee [8] reported that deregulation of E2F in the presence of p53 causes early entry into S phase and subsequent apoptosis in Rat-2 fibroblasts. Similarly, when the Rb-1 gene is knocked out, leaving E2F in its free, active form, mice are non-viable, with massive cell death from inappropriate S-phase entry and subsequent apoptosis [9]. Interestingly, proliferation may help to rescue cells from otherwise fatal signals. Spyridopolous et al. [10] reported that overexpression of E2F exerts a survival effect in proliferating endothelial cells and restores cell-cycle progression. It seems that inappropriate levels of unbound E2F force S-phase progression, which, if combined with p53 growth arrest signals, may confuse the system into inducing apoptosis. However, when cells are already dividing, overexpressed E2F may simply enhance proliferation. Thus the effect of E2F-1 is in balance with other competing stimuli in the cell cycle. Tumours occur when one or more factors are disrupted, as is the case when E2F-1 is removed in a knockout mouse. Surprisingly, hypoproliferation is not observed; instead we see apparently normal development with later occurrence of lymphomas, lung and reproductive tract tumours [11], demonstrating how important balance is in the cell cycle.

At the C-terminus of c-Myc are two nuclear-localization domains. The middle DNA-binding portion contains a leucine zipper, a helix–loop–helix motif and a group of basic amino acids used for DNA recognition. The N-terminus contains a transcriptional activation domain. This proto-oncogene is expressed in almost all proliferating cells, but is down-regulated in terminally differentiated cells. It has been shown to promote both proliferation and apoptosis. Its apoptotic action is p53-dependent in some cells but not in others [12]. As with E2F, overexpression of c-Myc in serum-deprived Rat-1 fibroblasts resulted in apoptosis in those with wild-type p53 expression [13], again illustrating the danger of conflicting cellular signals. Interestingly, overexpression of c-Myc also increased expression of the pro-apoptotic protein Bax [14]. It is possible that c-Myc may induce both death and growth pathways upon activation, with active inhibition of death being required for continued cell growth.

Expression level — Bcl-2:Bax ratio
Transcription and translation are not the only ways of controlling protein expression. Rapid degradation, conformational changes and proteolytic activation of inactive zymogens can all alter an enzyme's activity. Protein phosphorylation by kinases and translocation between intracellular compartments are other mechanisms of control.

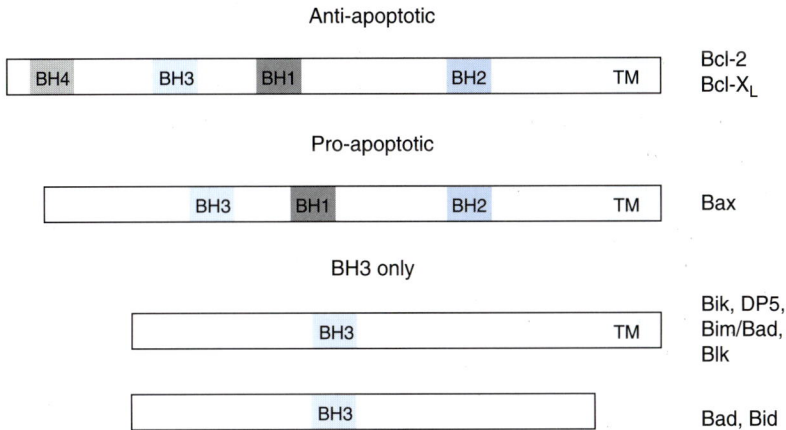

Figure 2. The Bcl-2 family
These cellular proteins regulate cellular fate through their interactions with family members and other cellular proteins such as Apaf-1 [15]. They share varying numbers of (BH) domains, and many contain transmembrane domains (TM) that anchor the protein to intracellular membranes. Here they interact with each other to form dimers; the particular configuration of the dimer is variable, with both homodimers and heterodimers being formed. Some of the BH3-only proteins are not bound to membranes and are thought to translocate to the membrane following activation.

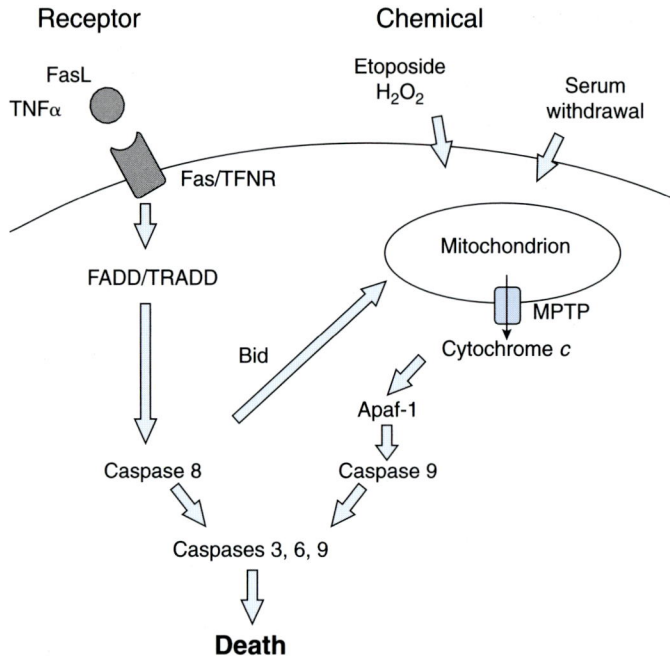

Figure 3. Receptor-mediated and chemically-induced death
Extra-cellular signals can be classified as either (a) ligands that bind to receptors or (b) chemicals that act directly on the cellular machinery. Ligands, such as Fas or TNF-α, bind to membrane receptors to activate intracellular pathways that eventually lead to cell death. Cytotoxic chemicals bypass the membrane and induce the caspase cascade by acting directly on mitochondria. The end result is the same however, with activation of the effector caspases and the induction of apoptosis. TNFR, TNF-α receptor; TRADD, TNFR-associated death domain protein.

Maintaining strict control over the life and death processes is of obvious importance to the cell. Two proteins from the same family, Bcl-2 and Bax, form a cellular 'rheostat' determining which pathway is chosen, i.e. whether the cell lives or dies, respectively. The cell survival protein Bcl-2 contains four conserved domains named BH1–BH4 (Bcl-2 homology domains) (Figure 2) and a hydrophobic C-terminal domain used for insertion into mitochondrial and endoplasmic reticulum (ER) membranes. Anti-apoptotic Bcl-2 and Bcl-X_L comprise only one of the three subsets in the Bcl-2 family. Pro-apoptotic proteins, Bax and Bid, and the BH3-only class (containing Bik, DP5, Bim/Bad and Blk) make up the other two.

Studies of the tertiary structure of Bcl-X_L have revealed a structure very similar to the pore-forming domain of diphtheria toxin. Bcl-X_L contains a pair of core hydrophobic helices that penetrate the lipid membrane and are shielded from the aqueous environment by five to seven amphipathic helices [15]. This is thought to be the key to their function, as they are known to be present in mitochondrial membranes, and the mitochondria play a central role in apoptosis (Figure 3).

The different members of the family have the ability to form heterodimers with each other. Bcl-2 (pro-survival) competes with Bax (pro-apoptotic) to prevent it forming death-inducing Bax:Bax homodimers. The Bcl-2:Bax ratio is therefore critical in determining sensitivity to apoptosis and eventual cell fate. Factors that alter this ratio in MIN6 and RINm5F pancreatic β-cell lines include serum deprivation and Ca^{2+} chelators such as EGTA and BAPTA [bis-(o-aminophenoxy)ethane-N,N,N',N'-tetra-acetic acid] [16]. Fatal exposure of MIN6 cells to 4 mM BAPTA or EGTA resulted in a steady decrease in Bcl-2 mRNA levels, whereas Bax mRNA increased in the 24 h following exposure. Removal of serum resulted in a decrease in Bcl-2 levels after 96 h, while Bax protein remained relatively constant. Further work showed that the changes in mRNA corresponded with changes in protein levels.

Environmental influences

Death induced by survival factor withdrawal or exposure to cytotoxic chemicals

Current research suggests that exposure to cytotoxic chemicals or serum withdrawal induces cell death by bypassing the cell's decision-making process and acting directly on mitochondria to release cytochrome c (Figure 3). This leads to the activation of apoptotic protease activating factor 1 and also caspase 3, the final executioner of cell death.

Hydrogen peroxide (H_2O_2) was used by Bladier et al. [17] to induce both apoptosis and senescence in primary human diploid fibroblasts, the outcome depending on the concentration of the H_2O_2. At 50–100 μM H_2O_2 there was an increased number of cells in G_1 and G_0 and a decreased rate of proliferation. This senescence-like state correlated with an enlarged and flattened morphology. Apoptosis was observed when the H_2O_2 concentration was increased to 300–400 μM. These results indicate a relationship between the concentration of H_2O_2 and the extent of the cellular damage. Interestingly, 200 μM H_2O_2 produced features of both apoptosis and senescence, suggesting that there is a linear relationship between H_2O_2 concentration and cell damage.

Receptor-initiated death

Receptor-initiated death is mediated by the tumour necrosis factor (TNF) family of cell-surface receptors, which are activated after exposure to Fas ligand or TNF-α (Figure 3). Binding of Fas leads to the activation of Fas-associated death domain (FADD) that in turn leads to caspase activation, as indicated by the ability of caspase inhibitors to prevent cell death in response to Fas. However, these inhibitors are ineffective against other forms of apoptotic stimuli such as DNA damage, activated oncogenes or Bak [18].

The intracellular pathways of chemically-induced and receptor-mediated death were studied by Sun et al. [19]. They used TNF-α and Fas to activate receptor-mediated pathways and etoposide to activate chemically-induced death. The caspase inhibitor, Z-VAD-FMK (benzyloxycarbonyl-Val-Ala-D,L-Asp-fluoromethylketone), was utilized to determine whether any divergence could be found between the two pathways.

The results were intriguing. In receptor-mediated death the inhibitor prevented the biochemical features of cell death, such as cytochrome c release, caspase 8 activation and cleavage of Bid. Sun et al. revealed an action upstream of the final commitment to die. In chemically-induced death the inhibitor again prevented, death but not cytochrome c release or the cleavage of Bid. Sun et al. took this to indicate that the inhibitor was acting at the level of preventing the manifestation of death (by acting on caspase 9) but not the decision to die, as indicated by the presence of cytochrome c and Bid. They concluded that caspases have a varying influence on the outcome of death depending on the type of stimulus. In receptor-mediated death these enzymes are inherent to the decision-making process, while chemically-induced death sees them relegated to the level of executioners of death only. This experiment strongly suggests that the method of inducing death is critical to whether the decision-making process is utilized or overridden. Viruses are a classic example of bypassing the intracellular death decision process.

Viruses — overriding cell choice

To survive inside a living cell without the cell recognizing the invasion and eliminating itself, a virus must block the cell death pathway. In this way, the virus overrides the decisions made by the cell to set its own agenda. There are three main cellular control points which viruses attack. Adenovirus and Epstein–Barr virus produce gene products E1B19K and BHRF1, respectively, which mimic the cell survival functions of Bcl-2. This transforms cells into continuously proliferating cells unable to induce death. Simian virus 40 and human papilloma virus modify p53 action and prevent its tumour-suppressor action, thus allowing dysregulated cellular proliferation. Cowpox and insect baculovirus produce CrmA and p35, which act as caspase inhibitors and prevent the execution of cell death.

The mechanism of death

The appearance of phosphatidylserine on the outside of the plasma membrane was assumed to indicate commitment to cell death. It may facilitate recognition by phagocytes, subsequent phagocytosis and degradation of the apoptotic cell. This can be visualized by the binding of fluorescein-labelled annexin V.

Hammill et al. [20] cross-linked B-cell lymphoma membrane immunoglobulin receptors to initiate cell death. They found that, although many cells bound annexin V, they were viable and could resume normal growth once the cross-linking was reversed. This demonstrates that the commitment to die had not yet taken place. Therefore, we can postulate that in these cells the timing of the critical decision is after annexin V binding but before caspase activation.

Caspases

Execution is carried out by activation of cysteinyl-aspartyl proteases, or caspases. Caspases are synthesized as inactive precursors and are activated via proteolytic cleavage by apoptotic protease-activating factor-1 (Apaf-1), FADD or other caspases (Figure 3). The active complex is a tetrameric enzyme made of two heterodimer caspase molecules.

Caspases 8 and 9, the initiator caspases, are interleukin-converting-enzyme-like proteases that are able to initiate the death cascade through recruiting other caspases. Caspase 8 is believed to be one of the first caspases cleaved by FADD following Fas activation. Once active it will activate other procaspase 8 as well as caspases 3, 6, and 7, thus creating a death cascade. The other initiator caspase, caspase 9, also targets these effector caspases, but in response to activation by Apaf-1. Harvey et al. [21] reported that activation of these caspases alone is necessary, but not sufficient, to cause death. They reported that Bcl-2 is able to rescue cells after activation of initiator caspases, but not once effector caspases had been activated. Therefore it is assumed that Bcl-2 acts at the boundary between initiator and effector caspases to prevent death.

Once the effector caspases are activated death is certain. There is an irreversible drive towards death that leads to DNA fragmentation, nuclear condensation, cytoskeletal collapse, genome degradation, formation of apoptotic bodies and cellular disintegration (Figure 4) [22].

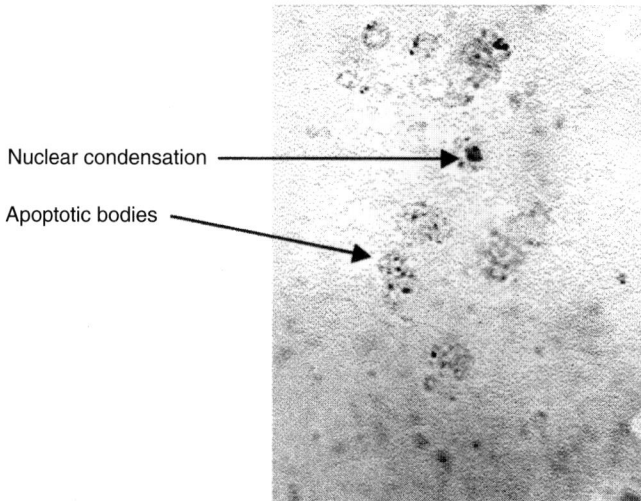

Figure 4. An apoptotic islet β cell with nuclear condensation and apoptotic bodies visualized using haematoxylin and eosin staining

Mitochondrial permeability transition pore (MPTP)

Apart from leading to a caspase death cascade, activation of the initiator caspases also opens the MPTP via cleavage of the Bcl-2 homologue, Bid, by caspase 8. Alternatively, the MPTP can be opened via a Bid-independent pathway following exposure to cytotoxic chemicals. Opening of the pore leads to further amplification of the cascade via release of cytochrome c and subsequent activation of Apaf-1 (Figure 3) [23].

Green and Amarante-Mendes [24] suggested that the opening of the MPTP is one of the irreversible points in the pathway leading to cell death. In addition, Scorrano et al. [25] have shown that the opening of the pore precedes, and is causally related to, apoptosis.

The pores are located in the mitochondrial wall at points where the inner and outer membranes make contact. Opening the pore directly by treatment with the cytotoxic chemicals oligomycin and atractyloside [24] leads primarily to dissipation of the proton gradient and uncoupling of the respiratory chain. Secondary events include generation of reactive oxygen species and cessation of ATP production. Apoptosis-inducing factor is released along with cytochrome c and contributes to the dramatic final stages of apoptosis. The entry of water through the open pore leads to mitochondrial swelling and rupture. There is some controversy over whether cytochrome c causes MPTP opening or is simply released as a result of it.

Future perspectives

Cell pathway choice is best viewed as an iterative process. Each injury or signal provokes a response that channels the cell in a particular direction. In the cell the presence or absence of control genes like *p53*, *E2F* or c-*myc* has a great influence on whether the next step is undertaken. Their relative expression levels in the cell vary over time and influence the likelihood of progression to the next step in the pathway. The effective manipulation of these control genes provides an entrance into the very control centre of life. Our increasing ability to regulate life and death of chosen cells opens up the possibility of new therapies in the fight against many diseases.

Discovering what causes a cell to choose to die is especially important for type 2 diabetes where there is a loss of functional β-cells within the diabetic pancreas. Under conditions of cellular stress caused by hyperglycaemia and hyperlipidaemia some people become diabetic while others do not. Elucidating the mechanism behind this selective β-cell death is one of the key areas for future research. Conversely, it is important to find out what causes a cell to become a tumour cell and choose continuous proliferation rather than death.

Much research is focusing on the exact point at which cell death becomes inevitable and the process irreversible. The same cell may well use different mechanisms in response to various stimuli. Alternatively, different cell types utilizing different components may use the same mechanism. At present the opening of the MPTP, the activation of effector caspases and the ratio of Bcl-2 to Bax all appear to be key points of cell commitment.

Summary

- *Cell death is one of several choices a cell faces in response to injury. The cell's inherent properties and its external environment determine which pathway is chosen.*
- *Interaction between transcription factors such as p53, E2F and c-Myc acts to finely tune the pathway selection process.*
- *Once the cell death pathway is initiated cell, survival proteins can stop it at different stages upstream of the activation of effector caspases.*
- *The exact point of no return along the cell death pathway is unknown, but is likely to vary between cells. It is unlikely to be a single step, but rather a short process leading to irreversibility of the pathway.*

References

1.	Hengartner, M.O., Ellis, R.E. & Horvitz, H.R. (1992) C. elegans gene ced-9 protects cells from pro-grammed cell death. *Nature (London)* **356**, 494–499
2.	Vaux, D.L., Weissman, I.L. & Kim, S.K. (1992) Prevention of programmed cell death in C. elegans by human bcl-2. *Science* **258**, 1955–1957
3.	Amundson, S.A., Myers, T.G. & Fornace, Jr, A.J. (1998) Roles for p53 in growth arrest and apop-tosis: putting the breaks on after genotoxic stress. *Oncogene* **17**, 3287–3299
4.	Thompson, B.E. (1998) The many roles of c-Myc in apoptosis. *Annu. Rev. Physiol.* **60**, 575–600
5.	Prives, C. & Hall, P.A. (1999) The p53 pathway. *J. Pathol.* **187**, 112–126
6.	Giaccia, A.J. & Kastan, M.B. (1998) The complexity of p53 modulation: emerging patterns from divergent signals. *Genes Dev.* **12**, 2973–2983
7.	Hengstschlager, M., Holzl, G. & Hengstschlager-Ottnad, E. (1999) Different regulation of c-Myc and E2F-1-induced apoptosis during the ongoing cell cycle. *Oncogene* **18**, 843–848
8.	Shan, B. & Lee, W.H. (1994) Deregulated expression of E2F-1 induces S-phase entry and leads to apoptosis. *Mol. Cell. Biol.* **14**, 8166–8176
9.	MacLeod, K.F., Hu, Y. & Jacks, T. (1996) Loss of Rb activates both p53-dependent and indepen-dent cell death pathways in the developing mouse nervous system. *EMBO J.* **15**, 6178–6188
10.	Spyridopolous, I., Principe, N., Krasinski, K.L., Shu-hua, X., Kearney, M., Magner, M., Isner, J.M. & Losordo, D.W. (1998) Restoration of E2F expression rescues vascular endothelial cells from TNFα-induced apoptosis. *Circulation* **98**, 2883–2890
11.	Field, S.J., Tsai, F.Y., Kuo, F., Zubiaga, A.M., Kaelin, Jr, W.G., Livingston, D.M., Orkin, S.H. & Greenberg, M.E. (1996) E2F-1 functions in mice to promote apoptosis and suppress proliferation. *Cell* **85**, 549–561
12.	Hoffman, B. & Liebermann, D.A. (1998) The proto-oncogene c-*myc* and apoptosis. *Oncogene* **17**, 3351–3357
13.	Evan, G.I., Wyllie, A.H., Gilbert, C.S., Littlewood, T.D., Land, H., Brooks, M., Waters, C.M., Penn, L.Z. & Hancock, D.C. (1992) Induction of apoptosis in fibroblasts by c-Myc protein. *Cell* **69**, 119–128
14.	Sakamuro, D., Eviner, V., Showe, L., White, E. & Predergast, G.C. (1995) c-Myc induces apoptosis in epithelial cells by both p53-dependent and p53-independent mechanisms. *Oncogene* **11**, 2411–2418
15.	Reed, J.C. (1998) Bcl-2 family proteins. *Oncogene* **17**, 3225–3236

16. Mizuno, N., Yoshitomi, H., Ishida, H., Kuromi, H., Kawaki, J., Seino, Y. & Seino, S. (1998) Altered bcl-2 and bax expression and intracellular calcium signalling in apoptosis of pancreatic cells and the impairment of glucose-induced insulin secretion. *Endocrinology* **139**, 1429–1439

17. Bladier, C., Wolvetang, E.J., Hutchinson, P., de Haan, J.B. & Kola, I. (1997) Response of a primary human fibroblast cell line to H_2O_2: senescence-like growth arrest or apoptosis? *Cell Growth Differ.* **8**, 589–598

18. McCarthy, N.J., Whyte, M.K., Gilbert, C.S. & Evan, G.I. (1997) Inhibition of Ced-3/ICE-related proteases does not prevent cell death induced by oncogenes, DNA damage or the Bcl-2 homologue Bak. *J. Cell Biol.* **136**, 215–227

19. Sun, X.-M., MacFarlane, M., Zhuang, J., Wolf, B.B., Green, D.R. & Cohen, G.M. (1999) Distinct chemical cascades are initiated in receptor-mediated and chemical-induced apoptosis. *J. Biol. Chem.* **274**, 5053–5060

20. Hammill, A.K., Uhr, J.W. & Scheuermann, R.H. (1999) Annexin V staining due to loss of membrane asymmetry can be reversible and precede commitment to apoptotic death. *Exp. Cell Res.* **251**, 16–21

21. Harvey, K.J., Blomquist, J.F. & Ucker, D.S. (1998) Commitment and effector phases of the physiological cell death pathway elucidated with respect to Bcl-2, caspase and cyclin-dependent kinase activities. *Mol. Cell. Biol.* **18**, 2912–2922

22. Dragovich, T., Rudin, C.M. & Thompson, C.B. (1998) Signal transduction pathways that regulate cell survival and cell death. *Oncogene* **17**, 3207–3213

23. Nunez, G., Benedict, M.A., Yuanming, H. & Inohara, N. (1998) Caspases: the proteases of the apoptotic pathway. *Oncogene* **17**, 3237–3245

24. Green, D.R. & Amarante-Mendes, G.P. (1998) The point of no return: mitochondria, caspases and the commitment to cell death. *Results Probl. Cell Differ.* **24**, 45–61

25. Scorrano, L., Petronilli, V., Lisa, F.D. & Bernardi, P. (1999) Commitment to apoptosis by GD3 ganglioside depends on opening of the mitochondrial permeability transition pore. *J. Biol. Chem.* **274**, 22581–22585

10

Future perspectives

Nick Hastie[1]

MRC Human Genetics Unit, Western General Hospital, Crewe Road, Edinburgh EH4 2XU, U.K.

Transcription — just one level of regulation

In this volume you will have enjoyed an excellent series of essays on gene regulation by leaders in their respective fields. The focus is on regulation at the transcriptional level. Although this is justified from the volume of work done to date, it is important to acknowledge that, in the flow of information from gene to protein, there are several intermediate stages at which regulation can and does take place. These include RNA splicing, mRNA transport, mRNA stability and translation. The excellent essay on translational control by Chris Proud (see Chapter 8 in this volume) demonstrates how this can offer many advantages over transcription as a control point. However, this is not the whole story, because post-translational modification of proteins is taking on greater and greater regulatory significance, be it phosphorylation, acetylation or methylation — all strategies for rapidly modifying protein function, often in response to extrinsic or intrinsic signals.

Although we are starting to get a glimpse of how these different processes work, we are only scratching the surface in many respects. In reality, it is clear that there is co-ordination between these different steps and, in the future, we will not be able to discuss control at transcription without reference to the upstream and downstream regulatory events. For example, the level of a transcription factor itself can be regulated at the translational level and as described in the essay by Melanie Lee and Stephen Goodbourn (see Chapter 6 in this volume); transcription factors may often be tethered in an inactive form in the cytoplasm and only become active and targeted to the nucleus after phosphorylation.

[1]*e-mail: n.hastie@hgu.mrc.ac.uk*

One exciting new development not covered in this book is the direct link between transcription and RNA processing. Over the past few years it has become evident that splicing is carried out co-transcriptionally. Furthermore, the C-terminal domain of RNA polymerase II binds directly to splice factors, thus recruiting the splicing apparatus to nascent transcripts [1]. The link goes beyond this, because this domain of the polymerase also interacts with capping and termination enzymes that allow the transcripts to be completed. Remarkably, the strength and type of promoter can also influence the pattern of alternative splicing [2].

Transcription factors and genetics

One way to assess the importance of transcription in regulation is to peruse the classical genetics literature. In their historic screen for developmental mutants in *Drosophila*, Nusslein-Volhard et al. (for review see [3]) identified a large number of mutants deficient in early patterning of the embryo and later segmentation. Previously, Lewis had identified the classic series of homeotic mutants associated with transformation of body structures from one type to another. We now know that a very significant proportion of the genes defective in these mutants encode transcription factors, the majority of these being proteins containing homeobox or zinc-finger motifs [3]. Furthermore, *Drosophila* developmental mutants led to the identification of the major intercellular signalling pathways involving the Wnt, Hedgehog and BMP (bone morphogenetic protein) families of regulatory proteins. These pathways, which are conserved in vertebrates, all work by activating transcription factors. To my knowledge, very few of the classical mutants involve proteins that regulate gene expression at the post-transcriptional level; however, there might be some bias here as DNA-binding proteins may have more recognizable signatures. Notable exceptions include the *Transformer* and *Transformer 2* gene products, RNA-binding proteins that control sexual differentiation in *Drosophila* by regulating the alternative splicing of the *Doublesex* transcript [4]. Also, the first vital step in development is to set up asymmetries in the egg. This is achieved in part by asymmetric localization of the mRNAs encoding the very transcription factors that regulate early pattern formation. In turn, it is specific mRNA-binding proteins that take the messengers for these transcription factors to their specific subcellular locations in the egg. So mRNA-binding proteins play essential roles in the very earliest stages of development.

Possibly the most profound demonstration of the importance of transcription factors comes from an assessment of their involvement in human genetic disease. Genes encoding transcription factors of many varieties have been found to be mutated in a large number of human genetic disorders [5]. These include many tissue-specific or developmental-stage-specific regulators that bind directly to DNA. For example, the genes encoding proteins with the paired DNA-binding motif, PAX6, PAX3 and PAX2, are mutated in aniridia, Waardenburg's syndrome and renal coloboma syndrome, respectively. Remarkably, these are all cases of haplo-insufficiency, where one copy of the gene is mutated and the remaining 50% dose is insufficient for normal development. Transcription factors of the steroid receptor class are mutated in a variety of diseases, including male breast cancer and vitamin D-resistant rickets [5]. Maturity-onset diabetes can arise through mutations in any one of several transcription factors that work together in pathways to control pancreas differentiation and the expression of insulin [5].

Arguably the most dramatic insights into the physiological and clinical significance of transcription factors have come through investigation into the genetic mechanisms underlying cancer. One of the biggest conceptual breakthroughs in cancer has been Knudson's two-step hypothesis and the consequent discovery of the tumour-suppressor genes. The most famous and commonly mutated tumour-suppressor genes, *p53* and *Rb1*, both work by regulating transcription — *p53* as a direct DNA-binding factor and *Rb1* as a co-repressor (see the essay by H.M. Chan et al.; Chapter 7 in this volume). Inherited mutations in these genes lead to very rare cancer syndromes. However, somatic mutations in these two genes occur in 50%, or more, of all of the common (sporadic) cancers. This new knowledge not only offers novel therapeutic avenues, but has also provided powerful insights into the regulation of the cell cycle and the response to DNA damage.

A vast range of transcription factors have been found to be activated by chromosome translocation in a variety of leukaemias [6]. Perhaps the most exciting recent development has been the finding that many chromatin modifiers, rather than DNA-binding regulators, are mutated in disease. As discussed in the essay by Alan Wolffe (see Chapter 4 in this volume), the biggest paradigm shift in the field of transcription has been the discovery that chromatin modification plays a key role in transcriptional regulation. Various modifying complexes, both those working as activators and those working as repressors, have been identified and a number of components of these complexes have been found to be mutated or rearranged in human disease, particularly in leukaemias and lymphomas.

Clearly it is exciting that these studies are revealing insights into the mechanisms of human disease and raising the prospect of new therapies. However, from a fundamental perspective it is pleasing that the human genetic findings validate the physiological importance of these factors and help both to support the biochemical studies and to place genes in cellular pathways.

What does the genome sequence tell us?

We are sitting at a major crossroads in biomedical science as the sequence of the human genome is nearing completion and the sequences of yeast, *Caenorhabditis elegans* and *Drosophila* genomes are complete. It is illuminating to dip into the sequence databases to assess the abundance of different classes of transcription factor gene within each genome and to see how this has changed with evolutionary complexity. In flies, genes encoding putative transcription factors with C_2H_2 zinc-finger domains are the most highly represented gene class in the genome, with 352 members [7]. There are also 130 genes encoding helix–loop–helix DNA-binding proteins, and 113 genes encoding homeobox proteins. It is interesting to note that, whereas there are more genes in *C. elegans* (18,424) than *Drosophila* (13,601), there are fewer representatives of each of these categories of putative transcription factor. Hence there are predicted to be only 138 zinc-finger proteins, 88 homeo-domain proteins and 46 helix–loop–helix proteins in *C. elegans*. So the increa-sed complexity of putative transcriptional regulators has co-evolved with increased biological complexity. However, another provocative observation is that the number of genes encoding proteins with RNA-recognition motifs has also increased from 55 in yeast, to 92 in worms, to 160 in flies. It will come as no surprise that we have no idea about the functions of the majority of these putative regulators, either transcriptional or post-transcriptional. We expect that the numbers of DNA- and RNA-binding factors in humans will have increased dramatically over the numbers found in invertebrates. Again this shows the scale of the challenge that awaits us.

As various genome sequences are reaching completion, the part played by alternative processing is gaining increased attention. It was a surprise to many that a fly has fewer genes than a worm, in spite of its more complex make-up and biology. One potential explanation for this is that there is more alternative splicing in flies, leading to greater complexity at the protein level. This possi-bility has been highlighted by the recent, remarkable finding that a gene encoding a fly axon-guidance receptor can produce potentially 38000 different proteins through a combination of differential splicing patterns [8]. Schmucker and co-workers speculate that this molecular diversity may contribute to the specificity of neuronal activity. This whole area of regulated alternative pro-cessing has received much less attention than transcriptional control, but this is likely to change in the near future.

Multifunctional regulators

It is also becoming more and more obvious that some proteins, which have been predicted on structural grounds to be transcription factors, may be multifunctional, acting at the post-transcriptional as well as the transcriptional level. For example, the famous bicoid DNA-binding transactivator also binds to the RNA transcripts of target genes, regulating translational efficiency [9]. The Tra-1 transcription factor, necessary for normal sexual development in *C. elegans*, not only regulates the transcription of the *Tra-2* gene but also binds to the Tra-2 transcript, regulating its transport into the cytoplasm [10]. Another remarkable example is the Wilms' tumour-suppressor gene, *WT1*, which encodes multiple protein isoforms with four DNA-binding zinc fingers. Whereas one of the major isoforms has high affinity for DNA and seems to act as a transcription factor, the other major isoform binds with higher affinity to RNA and interacts with splicing factors in the cell [11]. Such proteins may co-ordinate the processes of transcription and post-transcriptional events in particular developmental pathways. We must be careful in attributing a transcriptional function to a protein just because it has predicted DNA-binding motifs. Many DNA-binding proteins can also bind RNA with some sequence specificity and it may turn out that this is their primary role [12].

The future — 'massively parallel' and computational approaches

To understand fully development, physiology and the mechanisms underlying disease, we will have to work out the functions of all 60000–100000 human genes — how their products interact and talk to each other in complex networks. Scientists are just starting to use computational approaches to model complex interactions between developmental genes in *Drosophila* [13]. To deal with all this complexity, new 'massively parallel' approaches are being developed to study gene function. These include DNA chips or microarrays which make it possible to study the expression of tens of thousands of genes simultaneously. This approach can be used to identify the putative targets of a transcription factor. For example, in an elegant study, Soukas and co-workers [14] used microarrays to identify many of the mRNAs that were induced by the weight-controlling hormone, leptin, in adipose tissue. Among these were the mRNAs encoding some transcription factors already known to be induced in adipose tissue, along with the mRNAs for their suspected target genes. However, other transcription factors not known to be involved in the responses to leptin were also induced, with the same kinetics as several other genes which may turn out to be their targets. Thus, in one fell swoop, we are learning about the biological response to hormones and identifying transcriptional circuits.

All this has led to a massive rise in a new field of science: bioinformatics. Analysis of the change in expression of many thousands of genes simultane-ously has necessitated the development of sophisticated computing tools and massive databases. One laboratory may observe, for example, that ten specific genes are frequently expressed at higher levels in breast cancer cells compared with levels in normal mammary tissue. The scientists then access the databases and find that others have shown that the same set of genes is repressed by oestrogen, or perhaps down-regulated during normal mammary development. The more the databases expand, the more that can be learned from an experi-ment. Particularly exciting will be the information that is obtained on the many thousands of genes that encode so-called 'pioneer' proteins, i.e. those with no obvious biochemical function revealed by their gene sequences.

Ultimately, we will want to have a record of the expression patterns of all genes in time and space during development — a four-dimensional pattern of gene expression — and the only way to do this is electronically. Scientists in Edinburgh, at the MRC Human Genetics Unit and the University, have now started to develop a computerized, three-dimensional representation of a mouse at different stages of development (http://genex.hgu.mrc.ac.uk/). This is already a very powerful anatomy teaching tool. Into this atlas will be placed all the gene-expression patterns. It will then become possible to use computational approaches to identify clusters of co-regulated genes during development and to relate this to microarray data. This atlas will contribute enormously to our understanding of developmental processes, gene networks and disease mechanisms.

Summary

- *Transcription is coupled to splicing and other post-transcriptional processes.*
- *The importance of transcription factors in developmental biology and disease is underlined by genetic analysis in flies and humans.*
- *The genome project is identifying large numbers of novel transcription factors and RNA-binding proteins.*
- *Proteins may have multiple functions, acting at the transcriptional and post-transcriptional levels.*
- *The vast amount of novel biological information requires new, high-throughput approaches and bioinformatics.*

References

1. Bentley, D. (1999) Coupling RNA polymerase II transcription with pre-mRNA processing. *Curr. Opin. Cell Biol.* **11**, 347–351
2. Cramer, P., Cáceres, J.F., Cazalla, D., Kadener, S., Muro, A.F., Baralle, F.E. & Kornblihtt, A.R. (1999) Coupling of transcription with alternative splicing: RNA pol II promoters modulate SF2/ASF and 9G8 effects on an exonic splicing enhancer. *Mol. Cell* **4**, 251–258
3. Lawrence, P.A. (1992) *The Making of a Fly: the Genetics of Animal Design*, Blackwell Sciences Ltd, Oxford
4. Nagoshi, R.N. & Baker, B.S. (1990) Regulation of sex-specific RNA splicing at the *Drosophila doublesex* gene: *cis*-acting mutations in exon sequences alter sex-specific RNA splicing patterns. *Genes Dev.* **4**, 89–97
5. Engelkamp, D. & van Heyningen, V. (1996) Transcription factors in disease. *Curr. Opin. Genet. Dev.* **6**, 334–342
6. Rabbitts, T.H. (1994) Chromosomal translocations in human cancer. *Nature (London)* **372**, 143–149
7. Rubin, G.M., Yandell, M.D., Wortman, J.R., Gabor Miklos, G.L., Nelson, C.R., Hariharan, I.K., Fortini, M.E., Li, P.W., Apweiler, R., Fleischmann, W., et al. (2000) Comparative genomics of the eukaryotes. *Science* **287**, 2204–2215

8. Schmucker, D., Clemens, J.C., Shu, H., Worby, C.A., Xiao, J., Muda, M., Dixon, J.E. & Zipursky, S.L. (2000) Drosophila Dscam is an axon guidance receptor exhibiting extraordinary molecular diversity. *Cell* **101**, 671–684

9. Rivera-Pomar, R., Niessing, D., Schmidt-Ott, U., Gehring, W.J. & Jackle, H. (1996) RNA binding and translational suppression by bicoid. *Nature (London)* **379**, 746–749

10. Graves, L., Segal, S. & Goodwin, E.B. (1999) TRA-1 regulates the cellular distribution of the tra-2 mRNA in C. elegans. *Nature (London)* **399**, 802–805

11. Larsson, S.H., Charlieu, J.-P., Miyagawa, K., Engelkamp, D., Ross, A., van Heyningen, V. & Hastie, N.D. (1995) Subnuclear localization of WT1 in splicing or transcription factor domains is regulated by alternative splicing. *Cell* **81**, 391–401

12. Ladomery, M. (1997) Multifunctional proteins suggest connections between transcriptional and post-transcriptional processes. *Bioessays* **19**, 903–909

13. Von Dassow, G., Meir, E., Munro, E.M. & Odell, G.M. (2000) The segment polarity network is a robust developmental module. *Nature (London)* **406**, 188–192

14. Soukas, A., Cohen, P., Socci, N.D. & Friedman, J.M. (2000) Leptin-specific patterns of gene expression in white adipose tissue. *Genes Dev.* **14**, 963–980

Subject index

A

acetyltransferase, 50
activating transcription factor (ATF), 10, 41
adenylate cyclase, 72
annexin V, 116
Apaf, 117
apoptosis, 93, 109–119
apoptotic bodies, 117

B

bacteriophage λ, 7–10, 21
bacteriophage lambda (see bacteriophage λ)
Bak, 115
Bax, 111, 113
Bcl-2, 111
Bcl-X$_L$, 111
Bid, 116
bioinformatics, 125
BRG1–BAF complex, 54, 55

C

Caenorhabditis elegans, 110, 123
cAMP, 72
cAMP receptor protein (see *also* catabolite
    activator protein), 21, 22
cAMP-response-element-binding protein, 73
CAP (see catabolite activator protein)
caspase, 115, 116, 117
catabolite activator protein 5–7, 50
β-catenin, 84
cell cycle, 87–95, 112
chromatin, 45, 46, 51 61
chromatin remodelling, 91
λ cI protein, 24
c-Myc, 111

conformational change at promoter, 25
co-operative binding, 2, 4
CREB-binding-protein, 74
CRP (see cAMP receptor protein)
cyclin, 89, 90, 91
cyclin-dependent protein kinase, 89–92, 94
cytochrome *c*, 115, 116
CytR protein, 27

D

diacylglycerol, 74, 76
DNA fragmentation, 117
DNA methyltransferase, 60, 63–66
Dnmt (see DNA methyltransferase)
Drosophila, 105, 122, 123

E

E3 ubiquitin-conjugating enzyme, 83
E2 ubiquitin ligase, 83
E3 ubiquitin ligase, 83
EF2 protein 87, 111, 112
    interaction with pRb, 89, 90, 91, 93
    molecular structure, 89
    regulation of activity, 89–92
    transcriptional activation by, 92
embryogenesis, 105–107
enhansosome, 11
eukaryotic elongation factor, 100, 104
eukaryotic initiation factor, 99–101, 103, 104

F

Fas, 115
ferritin, 106, 107
filamentation response, 79